"One of the great strengths of this book is its rigorous and empirically rich account of Arctic cooperation from a role theoretical perspective. Sandra Engstrand's analysis brings together in-depth knowledge of Arctic state interaction, environmental and resource politics as well as foreign policy learning theory, thereby offering an excellent starting point to explain future cooperation and non-cooperation in this vital region."

**Sebastian Harnisch**, *Professor of Political Science,*
*Heidelberg University*

"Sandra Engstrand offers a refreshing view on Arctic relations, by identifying how Arctic states learn and play their roles. It is an important and awaited contribution to the Arctic governance literature. A must read for everyone interested in how states understand the Arctic."

**Svein Vigeland Rottem**, *Senior Research Fellow,*
*Fridtjof Nansen Institute*

"No serious scholar of environmental politics or international relations would deny that climate change in the polar regions has become a matter of urgent concern. If our present habits must change, then states must learn to embrace new behavioral norms. Sandra Engstrand uses role theory to give us new insight into the unfolding drama of Arctic Council negotiations, and new hope about the prospects for learning even as the geopolitics of environmental cooperation has become more fraught than ever."

**Paul Kowert**, *Associate Professor,*
*University of Massachusetts*

# Role Theory, Environmental Politics, and Learning in International Relations

In this book, Sandra Engstrand uses role theory to study learning processes in environmental policy negotiations in the Arctic Council.

Owing to rapid ice-melting in the Arctic region, and more accessible commercial opportunities, there is a greater need for environmental protection. However, large sections of the Arctic fall under state jurisdiction, often causing tensions to arise that prevent any cooperation from achieving fully efficient environmental protection. To enhance our understanding on how states learn about environmental norms, Engstrand examines negotiation processes on environmental protection for the prevention of Arctic marine oil spills and the reduction of short-lived climate pollutants. Through interviews with state representatives and through text analyses of nearly twenty years of meetings between Senior Arctic Officials from each of the eight Arctic states, Engstrand suggests that learning on environmental norms runs firstly through a learning of roles in international relations. She demonstrates how member states develop through self-reflection and by considering the expectation of others, concluding that states' wishes to pre-serve their social role in a group and to be perceived as Arctic 'cooperators' are drivers for a social education on environmental norms.

A timely and unmatched volume *Role Theory, Environmental Politics, and Learning in International Relations* will engage students and academic researchers in international relations, environmental governance, and Arctic politics.

**Sandra Engstrand** has a PhD in Political Science from Lund University. She has been teaching courses in International Relations at the Department of Global Political Studies at Malmö University, Sweden.

**Role Theory and International Relations**
Edited by Cameron G. Thies, Arizona State University, and Juliet Kaarbo, University of Edinburgh

The *Role Theory and International Relations Series* aspires to attract and publish the latest and best research integrating knowledge in the field of International Relations with role theory. This aspiration cuts across a wide swath of subfields, including foreign policy analysis, peace and security studies, international political economy, diplomatic studies, and international organization. While each of these subfields of study is presently organized as an 'island of theory,' this series intends to integrate their signature phenomena within a system of knowledge, a 'theory complex,' or an alliance among different subfields. This series showcases the ability of role theory to generate useful theoretical insights on its own or in combination with existing theories across these traditional subfields. Role theory's conceptual repertoire, its ability to span multiple levels of analyses, and the major meta-theoretical divides in the discipline position it to be an important integrative force in the study of international relations.

# Role Theory, Environmental Politics, and Learning in International Relations

The Case of the Arctic Region

**Sandra Engstrand**

NEW YORK AND LONDON

First published 2022
by Routledge
605 Third Avenue, New York, NY 10158

and by Routledge
2 Park Square, Milton Park, Abingdon, Oxon OX14 4RN

*Routledge is an imprint of the Taylor & Francis Group, an informa business*

*Library of Congress Cataloging-in-Publication Data*
A catalog record for this title has been requested

ISBN: 978-0-367-35138-0 (hbk)
ISBN: 978-1-032-01209-4 (pbk)
ISBN: 978-0-429-32999-9 (ebk)

Typeset in Times New Roman
by Newgen Publishing UK

**To August, Nils, and Ingemar**
*Three little brothers with the biggest of hearts*

# Contents

# Acknowledgments

This book is a revised version of my doctoral thesis. I'm grateful to Mikael Spång, Department of Global Political Studies at Malmö University, for all his insightful help and support over the years. I further wish to thank Rikard Bengtsson, Ole Elgström, Magnus Ericson, and Gunnhildur Lily Magnusdottir for having brought, in various ways, clarity to this work, as well as to all the anonymous reviewers for their valuable comments on how to move forward turning it into a book. Finally, I'm truly thankful to all the interviewees who participated in this study and generously shared their experiences and thoughts.

# Abbreviations

| | |
|---|---|
| AAC | Arctic Athabaskan Council |
| AC | Arctic Council |
| ACIA | Arctic Climate Impact Assessment |
| AEPS | Arctic Environmental Protection Strategy |
| AMAP | Arctic Monitoring and Assessment Programme |
| AORF | Arctic Offshore Regulators Forum |
| CAFF | Conservation of Arctic Flora and Fauna |
| COP | Conference of the Parties |
| EEZ | Exclusive Economic Zone |
| EGBCM | Expert Group on Black Carbon and Methane |
| EPPR | Emergency Prevention, Preparedness and Response |
| GEG | Global Environmental Governance |
| HoD | Head of Delegation |
| ICC | Inuit Circumpolar Council |
| IEA | International Energy Agency |
| IMO | International Maritime Organization |
| IPCC | Intergovernmental Panel on Climate Change |
| IR | International Relations |
| O&G | Oil and Gas (industry) |
| PAME | Protection of the Arctic Marine Environment |
| PP | Permanent Participant |
| RAIPON | Russian Association of Indigenous Peoples of the North |
| SAO | Senior Arctic Official |
| SLCP/-F | Short-Lived Climate Pollutants/Forcers |
| TFBCM | Task Force for Action on Black Carbon and Methane |
| TFOPP | Task Force on Arctic Marine Oil Pollution Prevention |
| UNCLOS | United Nations Convention on the Law of the Sea |
| UNFCCC | UN Framework Convention on Climate Change |
| WG | Working Group |
| WWF | World Wildlife Fund |

# 1 Introduction

## Learning in Arctic Environmental Cooperation

## Introduction

The Arctic region is in desperate need of enhanced environmental governance, or, in the lingua of this book, of international learning. Over
the past fifty years, the Arctic temperature has rapidly increased, with
an average warming of two to three times higher than the global average
(IPCC, 2018). The ice and frozen tundra have long symbolized this region,
but now these symbols of the Arctic are in decline. Climate change has
caused the permafrost to thaw, the sea ice to melt, and glaciers to increasingly calve into the Arctic Ocean. As early as 2030, or soon thereafter, it
is expected that most of the Arctic will be free of ice during late summer.
The Arctic Monitoring and Assessment Programme (AMAP) – a scientific working group (WG) under Arctic Council – has described the Arctic
to undergo a transformation, stretching beyond visual ice-melting to its
aftereffects of ecological changes and disruptions (AMAP, 2017). Indeed,
Arctic warming has turned not only into a regional catastrophe by severely
affecting many indigenous communities with traditional ways of Arctic
living, but also as a catastrophe for the world at large because an Arctic
region lost to uncontrolled ice-melting and thawing permafrost will have
the world soon to follow.[1]

It takes an international learning on environmental norms to have this
negative and downward warming trend dampened. Following J. Knopf
(2003), such learning should be shared amongst actors and it should be progressive, adhering to norms benefitting nature. History has before showed
international society capable of curving critical and downward trend in
emissions and species extinction, for instance, by implementing regulations
on ozone depleting substances and whale hunting alike, to mention a few.
On the contrary, history has also revealed international society being, for
too long, impervious to scientists' repeated message of the causal linkages
between anthropogenic greenhouse gas emissions and global warming.
The Arctic region is not only vulnerable to global warming, but its nature
and ecosystem are particularly fragile and vulnerable to marine oil spills
too. The continuous extraction of Arctic oil and gas, both on and offshore,
is a telling example of commercial interests yet to face a counterpart – i.e.,

an environmental norm – strong enough to have the state support withdrawn from environmentally risky activities.

How then do international relations learn of environmental values in such a way that what previously was an accepted behavior suddenly gets dated and perceived as harmful and obsolete? Obviously, the key would be to know about the problem, to have access to knowledge and scientific facts. But knowledge alone does not bring any learning to life – if so, many of the problems and horrors in the world would since long have been gone. It also takes another, more subtle thing; it takes the insight of the problem being precisely that – a problem. This book engages in a broader discussion on how a learning of environmental norms may come about, though it may be difficult to achieve. By applying a role theoretical perspective on International Relations (IR) in general and Arctic cooperation in particular, it is set to offer one piece to the puzzle on why knowledge and scientific facts alone seldom is enough to swiftly put an environmentally harmful activity on halt.

More specifically, the book approaches cooperation as a social learning process and argues that any learning on environmental norms firstly runs through a learning of roles in international relations. A state's role, in turn, is suggested to be a stable although flexible construction, informed by situated understandings. At the book's disposal are two negotiation processes that have been carried out by the Arctic Council in the years 2013–2015: marine oil spill prevention and reduction of short-lived climate pollutants, respectively. These case studies shed light on why there would be no such thing as an *instant* learning on environmental protection; instead it is suggested that a profound learning of environmental values by necessity must take time, time enough to allow for role-playing states to have their roles adapted to a new and somewhat changed script.

### *Approaching Learning Processes*

In IR, it has for long been a well-known fact that laws and regulations aiming for environmental protection generally battle in headwind. Difficulties pertaining to, for instance, competitiveness and risks of free-riding do often put strains on what would be cooperatively possible to achieve. In the Arctic, preventive measures on environmental pollution are furthermore held back due to strong economic incentives, where the region's vast resources of oil and gas, fish, minerals, and tourism all entail commercial promises that give rise to activities not always compatible with keeping pollution levels at a minimum for the sake of the environment. In various degrees, the Arctic states are all forced to relate to such economically informed interests, while nurturing the desire to have the region environmentally preserved and protected. Obviously, all environmental governance, Arctic and other, would be the result of interest leveling.

Rather than focusing on Global Environmental Governance (GEG) as a result, this book takes as a point of departure in the *process* of getting there, why do we – the Arctic states specifically and international actors in general – end up where we end up, when wrapping a negotiation session up? As the title of this book suggests, role theory, environmental politics, and learning are of interrelated relevance to the Arctic region. Indeed, the route towards greater Arctic environmental protection is argued to run through states' learning about their (social) roles: the wish to obtain a specific position in a group is as much – or even more – of a driver for state learning of environmental norms than the actual environmental norm per se. The book sets out to investigate the following question: *How do states learn of environmental norms?* This question is broken down also in smaller areas of interests, providing different types of input regarding the following: how is the intertwinement between energy and environment expressed in the Arctic Council interaction, how is learning manifested within such interactive processes, and how do expectations serve to either encourage or constrain learning? Role theory is deemed well-suited to engage with the overarching question of how states learn of environmental norms, since the role concept acknowledges actors to be both socially reflective and sensitive to behavioral prescriptions, while adhering to an inner core of Self which cannot easily be disturbed. Indeed, the very act of role-playing is, just like learning, a process. From this perspective, the link between roles and learning could be manifested as role change (Aggestam, 2006; Harnisch, Frank and Maull, 2011; Harnish, 2012). To quote the example of a theater, roles are stable over time and they follow a (national) script, but they are also attentive to the director, co-actors, and audience, and therefore hold the potential for change.

"You'll try to get an agreement on things where, you think, it will be possible to agree," a state representative with extensive experience of the Arctic Council explained, "because, if you get *one* agreement, you may be able to get another, which then will increase the amount of understanding between interested nations." The words are dedicated to what the general negotiation strategy of the Arctic state would be on a topic like oil spill prevention. Not only do they point toward the importance of a micro perspective to grant access to learning processes, but correlate with how this book foresees that negotiations and cooperation on Arctic environmental protection can be made progressive, as a stepwise and contextually anchored process, where actors little by little have their roles and perceptions altered. Indeed, to have actors to meet, and after having met, to meet again and again seems to be the socializing recipe benefitting the Arctic environment.

This chapter discusses how roles and role-playing is a good way to approach state behavior as pervious to social interaction. It also discusses norms as mattering in IR, including the norm of environmental responsibility. It ends with a discussion of the micro perspective as constituting an analytical bridge between norms and behavior, thereby rendering learning in IR less abstract. Discussions carried out at this level is set to contribute to the understanding of how global environmental protection in the end is a learning process originating in reflections, perceptions, understandings, and expectations found in the 'small.' The closing section of this chapter presents the Arctic case studies and methods relied on for subsequent discussions on learning processes.

## Arctic Role-Playing

In recent years, the Arctic has gained increased attention in media and news reports, bringing to fore an urgent need for strong Arctic cooperation on environmental protection. Since 1996 the Arctic Council (hereinafter the AC), an intergovernmental organization, has gathered states and indigenous associations to interact in what today represents perhaps the most stable and encompassing institution for cooperation in the Arctic (Dodds, 2013; Nord, 2016a; 2016b). The eight Arctic member states are: Canada, Denmark, Finland, Iceland, Norway, Russia, Sweden, and the U.S. According to the founding statue, cooperation should be promoted "on common Arctic issues," in particular on issues of sustainable development and environmental protection in the Arctic, while matters pertaining to militarily security are explicitly excluded (*Ottawa Declaration*, Art. 1a). Although it was formally established in the mid 1990s, the idea to bring the Arctic states close to one another through Arctic cooperation dates to the Cold War era. In a now famous speech, Mikhail Gorbachev declared the Arctic region to represent the opportunity of a peace zone, where also the environment would have much to gain from the burying of nuclear arms and agony. In 1989, on the initiative of Finland, the eight Arctic states agreed to the Arctic Environmental Protection Strategy (AEPS), thereby stipulating environmental cooperation as the way forward for not only the region, but one may presume, Arctic relations alike.

The Arctic region has been described as a different, even unique, regionalization process due to its attempt to enhance Arctic governance while yet being constituted by competing interests and discourses. The Arctic Council has similarly been considered a unique model for peace- and stability-building (Albert and Vasilache, 2017; Brigham et al., 2016). Contributing to this view is the acknowledgment of the importance of having also indigenous representation to inform the decision-making process, moving the intergovernmental cooperation beyond state perspectives strictly.

In the last decade, international cooperation in general seems to be waning. Multilateralism has suffered from setbacks. For instance, in Europe, the EU has lost one of its members to Brexit and temporary border controls have been reintroduced in the otherwise border-free Schengen area as a response to first migration and the coronavirus pandemic. In the U.S., the Paris Agreement has been withdrawn, and steps have been taken to a formal withdrawal from also the World Health Organization. Neither has the Arctic cooperation turned out immune to global events. A good example is the eruption of the Crimea conflict in 2014, which, although not occurring in the Arctic region per se, yet created serious concerns over Russia's military behavior (remilitarization) heading also the

Arctic. However, whereas relations with Russia were damaged, both bi- and multilateral ones, its effect on Arctic Council cooperation was moderate (Burke, 2019). Here, in the AC, cooperation has withstood. Common Arctic interests, in addition to an identified need to externally show themselves united as Arctic actors to be able to address and deliver on internal problem, have been suggested as what spurs continuous Arctic cooperation (ibid.:173).

The world is not getting less interdependent but, if anything, more interdependent. With an ongoing regional transformation of the Arctic, the pressure on the Arctic Council to find common solutions increases. Cooperation represents an opportunity to forestall what is happening, or, as many Arctic Council member representatives put it, cooperation is an opportunity to act before the region gets too crowded, too filled with activities (interviews, 2014–2015). By the same token, Arctic cooperation ensures regional control.

### Arctic Governance

The Arctic Council has been studied quite vigorously, especially during the last few years, and since before, there exists much literature how to move regional governance forward through the Arctic Council (Axworthy, Koivurova and Hasanat, 2012; Dodds, 2013; Nord, 2016a; 2016b). It has been suggested that there is a need for a more all-encompassing Arctic environmental treaty or framework that to a greater extent takes into consideration the transboundary character of eco-systems (Koivurova and Molenaar, 2010), whereas others have investigated the possible roles that the AC can take on as the region is now being faced with increased activity and growing environmental challenges (Nilsson, 2012; Pedersen, 2012; Wilson, 2016). It has furthermore been proposed to be a potential agent of change, which despite institutional challenges works to increase understanding and awareness of common Arctic issues (Barry et al., 2020). In addition, others have linked the Arctic Council success to the extent that it manages to affect or be backed by other (international) institutions (cf. Humrich, 2013; Young, 2009; 2016), whereas others have such success linked to the participation of indigenous peoples in norm- and decision-making (Koivurova and Heinämäki, 2006; Poto and Fornabaio, 2017). In a region where geopolitical stakes are rising, the environmental field has been foreseen to represent a cooperative entry into enhanced state security in the Arctic (Stokke, 2014).

Also this book is interested in the issue of good environmental governance, but it takes a somewhat different entry; instead of entering discussions through regimes and political representation, this book argues the prospects for cooperation and governance to go through cognitive mechanisms. Focus is thereby shifted from environmental issues per se to social interaction and the possible impact this will have on states' readiness to accept or even internalize environmental norms. A *role theoretical* perspective on Arctic relations has, to the author's knowledge, not been applied before, neither has Arctic negotiation been studied from a micro perspective as to why there would be a theoretical value in showing role theory as being well suited to localize, identify and reason about the Arctic state learning. The empirical value would be an enhanced knowledge of the interactive foundation for what is going in the Arctic.

The role is a tool for states making social interaction comprehendible. Stephen G. Walker defines roles as "repertoires of behaviour, inferred from others' expectations and one's own conceptions, selected at least partly in response to cues and demands" (Walker, 1992:23, in Thies, 2010:6336). Sebastian Harnisch narrows it down to "social positions (as well as a socially recognized category of actors) constituted by ego and alter expectations regarding the purpose of an actor in an organized group" (Harnisch, 2015:5, see also Harnisch, 2011). A description of roles as 'social positions' sheds light on the relational dimension of roles, which ought to make states more inclined to act responsively to the context. With role theory's origin in sociology, social psychology, and anthropology, it draws attention to the cognitive dimensions of IR, transforming state interaction into social acts, with the potential of having international relations re-written to pay greater attention to issues of good environmental governance.

Arctic role-playing does not exist as only a theoretical framing but is an experienced reality among negotiators and state delegates. One person, acting as a leader in a negotiation on emission reductions of black carbon, explained; "You, acting on behalf of your state, always take on a role," and continued by noting the varying level of commitment among state delegations through their role: "Some have not been so active, others have been more active, and the reason thereof is a result of thousands of things. That is just how it is. It is, how should I say, it is natural." Indeed, that negotiations develop in different ways is an effect of many things, for instance, time and space, past experiences and future expectations, national capabilities and international relationships. The list could go on but when applying role theory, it gets narrowed down; instead of a 'thousands of things,' what is left is 'expectations.' Expectations serve as an umbrella for all those different ideas and perceptions a state has to consider and balance when interacting with others – like interests, culture, and norms, and it includes both internal and external expectations; from the domestic context, from fellow negotiators, and from norms encircling international interaction. Role theory thus offers a perspective on IR where state behavior is seen less as state of the art, or as a static condition given by nature, and more so as social acts of role-playing.

### *Why Role-Playing Matters for the Arctic*

That the Arctic is changing is a fact. However, not every aspect of Arctic warming constitutes a threat to the Arctic region; the regional transformation it causes also brings commercial opportunities. This development is often referred to as the 'Arctic paradox': what is severely damaging the region from one perspective is bringing prosperity from another. Because, as the ice-edge withdraws north, attractive resources like oil, gas, minerals, and fish become more accessible. Possibilities for cost- and time-saving maritime transportations grow, not least through the Northern Sea Route which is expected to cut transportation time between China and Northern Europe by about 40 percent, compared to present route via the Suez Canal, thereby saving shippers many days in sailing time and reducing the amount of fuel used. At the same time, the more human activity there is in the Arctic the higher the risk for accidents, and likewise, for every additional extracted barrel of oil emissions of greenhouse gases and climate pollutants will follow.

In 2008, the U.S. Department of Interior published an assessment of remaining oil and gas resources in the world, where the Arctic was estimated to store about 20–25 percent of all yet undiscovered hydrocarbons reserves in its sediments. The year before, in 2007, a Russian expedition caused political controversy when planting a titanium Russian flag on the seabed, 4,261 meters beneath the northern point of the world. Although the activity was explained to be a scientific investigation, the flag-planting episode, in a context where commercial Arctic opportunities were increasingly lying at least potentially bare, displayed Arctic geopolitics on the rise. Predictions of Arctic scrambles were voiced, of cooperative meltdowns and of territorial conflicts, rendering the Arctic future an arena for rivalry and competition (see Borgerson, 2008; Posner, 2007).

However, more than a decade later, territorial conflicts and resource scrambles are still visibly lacking. Instead the Arctic states have teamed up behind Arctic Council cooperation. One of the Arctic Council participants interviewed for this book, having insight form various AC-related work, explain such cooperative development in the following way:

> It is not about putting flags in the ocean…you know, the journalists they always talk about the same thing; they talk about the explorer putting a Russian flag on the seafloor. This is not why the Arctic Council is important. The Arctic Council became important because the region is important for the earth system. And we start to understand that.

Yet, to acknowledge the environmental value of a region does not exclude the presence of minor conflicts and political resistance. As a matter of fact, in an Arctic region characterized by climatological changes, role-playing is as relevant as ever. It is relevant since the roles taken on by states in AC cooperation need to consider environmental responsibility should the Arctic warming trend not turn out irreversible. At the same time, most of these roles reveal discordant interests, making up a small 'Arctic paradox' of their own as regards holding state interests of both commercial and environmental character. In practice, this means that Arctic role-playing – that is, how the Arctic states decide to play their roles – indeed is relevant for the way future Arctic eco-systems turn out and for the approximately 500,000 indigenous people who live in the Arctic and who find their traditional way of living being impeded (AMAP, 2017). Arctic role-playing is, however, not only important to the Arctic region but the world at large; scientifically, there is a causal linkage between Arctic warming and global climatic system. Arctic warming is affecting the Northern hemisphere and its distant southern counterpart alike; it provides for climate feedback by reflecting less solar energy causing more ice to melt, by raising sea levels and by altering ocean circulation patterns, and by furthering the release of greenhouse gases as the permafrost thaws (Arctic Council, 2004). Therefore, depending on how the Arctic states choose to play their roles, Arctic role-playing matters for the way future evolves.

For about twenty-five years, Arctic state cooperation has revolved around goals of sustainable development and environmental protection in a rather peaceful, good, and lesser-tensed manner.[2] For instance, when Finland in 2017 shouldered the chairmanship of the Arctic Council, the country's Foreign Minister Timo Soini summarized the last few decades of cooperation by referring to the Arctic as of common interest: "The Arctic Council has promoted common Arctic interests worldwide." He also wished for the AC to stay committed working together for the sake of the Arctic people and environment (AC Ministerial statement by Finland, 2017). In Ministerial statements one after another, the Arctic states furthermore describe the Arctic Council cooperation to build on trust, where members should pay each other respect, and where a good 'discussion' climate is characterizing meetings.

For the Arctic region to develop soundly, it is paramount that the Arctic states will continue such a cooperative path. So far, this also seems the most likely outcome, given all the cooperative components that are included in the roles of the Arctic states. In Chapter 4 of this book, the behavioral pattern of each Arctic state is identified as heavily constructed around that of being an 'Arctic cooperator,' although with a national twist. The roles are: Canada – *Protector*, Denmark – *Pragmatic (coastal state) kid-brother*, Finland – *Reserved team player*, Iceland – *Follower with ambitions*, Norway – *Restless know-how leader*, Russia – *Responsive informant*, Sweden – *Teacher on demand*, and U.S. – *Innate leader*. These roles and their content will be returned to in subsequent chapters, and what suffices for now is that Arctic role-playing does not reveal Arctic relations to one-sidedly put economic gains, energy security, and business opportunities prior to Arctic environmental protection. In general, the eight Arctic states reveal few signs of acting in a cooperatively contra-productive manner: eight states engage in cooperation and eight states *do* cooperate.

Now, to find cooperation in precisely that, cooperation that is, should come as hardly a surprise. Still, an extensive survey on attitudes to international cooperation asked leaders in six different countries, including Russia and the U.S., and found an overall strong support – stronger than expected – for international cooperation, including within the environmental field (Scholte, Tallberg and Verhaegen, 2019). According to the same survey, democratic processes aiming for efficient problem solving are important components of international cooperation (ibid.). The AC confirms this result; here, cooperative processes are understood to be democratic, and they aim to solve common challenges (interviews, 2014–2015). This, in combination with the fact that the Arctic states seem to care for their reputation as 'good cooperators,' increases the likelihood for Arctic cooperation to be receptive also to the norm of environmental responsibility. If so, this could tentatively allow for restrictions in material and economic Arctic interests, resulting from role changes and Arctic state learning.

## A Greener IR, a Greener Arctic?

A prerequisite for a learning of environmental norms in IR is a belief in progression, for instance, through GEG. This field aims for, as Speth and Haas (2006) describe it, "planetary stewardship" by finding ways to reduce and alleviate the pressure put on the environment by the last century's substantial population growth and growth in economic activity (ibid.). Since the global order has also become a multicentric one – affected by many actors, such as states, non-state and private actors alike – GEG is argued to represent a transformation of world politics (Pattberg, 2012:98). The interest is linked to regimes and how they transform internal relations, and how regimes could be rendered more efficient (Humrich, 2013; Nilsson and Koivurova, 2016; Stokke, 2013; Young, 2008). Oran Young, leading scholar within the field of Arctic governance, describes what is needed for such efficiency in the following way:

> What is needed, for best results, are sets of rights, rules, and decision-making procedures that are not only accepted by those subject to them as proper or legitimate in principle but that also become sufficiently embedded or entrenched that key players participate in the resulting practices without thinking about the pros and cons of doing so each time they act in a manner that conforms to the relevant rights and rules.
>
> (Young, 2008: 21–22)

This description calls for a socialization effect, if GEG should have an imprint on international relations. Indeed, there are indications of such socialization occurring today, shaped as an international greening process. Following Robert Falkner (2012), this "greening of international society" implies that, much like how ideas such as sovereignty, international law, and the market have always guided international relations, so too has global environmental responsibility become a norm to consider. Clapp and Dauvergne describe it as "[t]he history of global environmental politics is inextricably tied to contest of ideas: battles of worldviews and discourses" (2011:45). These ideas are sent back and forth between different regimes, creating horizontal interlinkages between them that become "sites for collusion or contestation over these broader [governance] norms" (Zelli, Gupta and van Asselt, 2012:176). And so, over time, global environmental responsibility has gained an undisputed position within the global setting on norms where actors, like representatives from states and businesses, today are forced to assess, and possibly review, their interests and behavior against an international playing-field that in an increasing scale requires a responsible dealing with nature. Of course, speaking about IR, actors are not in an absolute sense being *forced*, but they are, however, pushed to act responsible to be perceived as legitimate: to gravely go against environmental science, in word or deed, generally means bad business and bad politics. In other words, actors are socialized into taking the environment into consideration when doing business.

An illustration of this development and how perceptions of legitimacy affect the rationale for cooperation is provided for by Wood-Donnelly (2016). In her discussion of lessons to be learnt for the contemporary development of hydrocarbon in the Arctic, comparisons are made with the Arctic whaling industry over a century ago, where the demand for oil (as lubricants) from the sperm whale did not cease until it was internationally recognized as an endangered species, causing a social pressure that made whale oil politically – and thus commercially – unviable as a product (ibid.:136). A current Arctic example which signals an environmental responsibility on expansion is the rather recent cooperation around an issue such as prevention of oil spill. Indeed, besides the obvious environmental gains of preventing marine oil spills, and the little less obvious – yet accurate – economic benefits of such prevention (Hasselström et al., 2017), it is today widely acknowledged that a major oil spill in the Arctic would severely damage the petroleum industry at large, no matter where in the Arctic the accident occurred; further energy activities would simply be met with too much resistance from the environmental movement. Back in history, 40–50 years ago when the extraction of hydrocarbons like oil and natural gas on a general basis first begun in the offshore Arctic, neither oil spill preparedness nor oil spill prevention were subject for international governance. The same is true for short-lived climate pollutants; not until recently has black carbon-producing activities, such as gas flaring or to fuel vessels with bunker oil, been considered problematic from an environmental perspective. Thus, things – attitudes, needs, and behavior – do change, and what once was perceived a legitimate action may in a new discourse as well as reality find itself to be illegitimately perceived.

### The Restraining Effect of Sovereignty

Despite the increase of demands for global governance over the years, there has only been a marginal increase in state capabilities regarding cooperation (Franchini Viola and Barros-Platiau, 2017). How come international cooperation, generally, is one step behind economic growth and resource exploitation? Part of the answer could be derived from sovereignty, which is a norm considered and expressed in parallel to environmental responsibility.[3] Also the leading authority within the environmental field, the United Nations Environment Programme (UNEP), struggles with this norm: whereas it describes itself to set the environmental agenda, to inspire actors to care for the environment, and to promote coherent implementation of environmental issues, it is yet a member state–led organization operating under the UN umbrella which places the protection of state sovereignty among those core tenets that ensure world peace.

The things the UNEP – and GEG in general – can do to push for enhanced environmental responsibility are unquestionable constrained by state sovereignty, which often prioritizes differently and applies other perspectives on international relations. Any international environmental cooperation therefore needs to find the right – as in politically acceptable – balance between these two norms and their prescriptions. The Arctic Council illustrates this 'battle' by its steady production of Arctic-specific knowledge pushing the frontiers regarding environmental responsibility forward, while at the same time, being criticized for being politically slow-paced, with weak political commitments and with few requirements placed upon states regarding implementation (Dubois and Tesar, 2014; Chuffart, 2019; interviews, 2014–2015).

Although a bit of an oversimplification, international environmental negotiations could be envisioned as a struggle between, on the one hand, connections and values made visible by (environmental) science, and on the other the realization that any change requested for does come with a significant political price. The acceptance of (what to view as severe) environmental degradation is a *choice* resting upon values; why it is politics, not science, that determines whether environmental measures should be taken or not (Sörlin, 2014; Wormbs and Sörlin, 2013). Similarly, whereas the AC statute stipulates sustainable development to be a key objective, sustainability itself would be a political concept that contains competing visions of the future, manifested as a discursive struggle regarding future rights and resources in the Arctic (Gad and Strandsbjerg, 2019). On a general level, struggles on values suggest that an environmental issue with high political controversy attached would seemingly not allow science to inform state decisions in any substantial way (see Rottem, 2017:250). This has since before been confirmed in the Arctic case where Langhelle, Blindheim and Øygarden (2008) have showed Arctic state cooperation to frame oil and gas as of superior importance; one has talked about the impact of Arctic climate change on oil and gas, but not the reverse.

Still, years have passed since this study, and today, possibly, climate change is no longer as controversial as before, not even in the Arctic. Not that many years ago, for instance, Norway, eagerly backed up by Russia but also Canada and Denmark, expressed deep unsatisfaction with how they understood oil extraction to repeatedly be presented in negative terms in AC discussions and cooperation – as simply environmental risks, rather than acknowledging also the positive effects its extraction may bring for economic development (SAO meeting Fairbanks, 2016). However, as will be discussed in coming chapters, several states still obstruct an explicit connection being made between Arctic warming and the burning of fossil fuels, and in AC cooperation, furthermore, oil and gas production remains an issue that falls under national sovereignty.

As a matter of fact, resistance is found everywhere, but in IR with less political controversy attached: whenever states fear that their interests are in danger they simply attempt to resist. Any learning of environmental norms therefore needs to, in some sense, cut back on resistance and instead embrace (the promise of) change. To understand how resistance affects the consensus-reaching process, the three interrelated variables of saliency, credibility, and legitimacy are good indicators of whether a scientific environmental assessment is likely to (be allowed to) influence policy or not: is the environmental norm relevant to the specific context, and is it understood to be scientifically relevant as well as politically acceptable? (Clark, Mitchell and Cash, 2006:12; see also Stokke, 2013; Nilsson, 2007). This assessment process is not restricted to inter-state bargaining based on national interests but is as well informed by environmental science and socialization, i.e., there are *different sources to learning*. Science and hard facts are one source, whereas social cues, demands, and expectations would be another, which is the predominant interest of this book.

## State Learning

In international relations, the actors who sign declarations are states. It is furthermore the states that make environmental commitments, states that implement them, and states that report on how their commitments all progress.[4] It is also the states that grant permission for industrial establishments like oil exploration or maritime transportations in and across the Arctic. Therefore, states are in a way the targets for learning in international relations. This book applies a concept such as 'state learning' to refer to and conceptually approach learning as a process. As such, the approach is similar to Farkas' (1998) who argues state learning to be the intervening process between international change (the cause) and policy change (the effect). However, in this notion of state learning, the research interest is located to the understanding of how rational individuals – acting on behalf of the state – arrive at foreign policy decisions in a somewhat materially changed world (Farkas, 1998). This book, to the contrary, views *social interaction* – norms and expectations – as what constitutes change, thus acting as driver for learning processes. The effect of state learning is not uttermost argued to be about empirics or policy changes within the environmental field – even if this may be the concrete and long-term objective – but to be about those preceding social learning processes that also entail reflections of their social roles.

An analysis of state learning presupposes the involvement of several analytical levels. For instance, a macro-level study of learning is often interested in the efficiency of GEG, and in the Arctic case specifically, focus has often been dedicated to questions such as how implicit and explicit rules and regulations are working to transform international politics, having the issue of efficiency in the back of the mind (Humrich, 2013; Nilsson and Koivurova, 2016; Stokke, 2013, Young, 2008). At the meso-level, learning is expressed by the way that states behave – perform their roles, whereas a micro-level analysis of learning manifests roles as a result of cognitive processes. Analytical levels are thus understood to 'frame space' into expressions of different dimensions of social interaction (Onuf, 1995); rather than depicting levels as hierarchically ordered, they add sources of explanation. For an area of investigation such as state learning, the micro level adds the much important component of making meaning systems. Meaning making – how state representatives understand and relate to the processes of which they are part – is a bearing component when attempting to understand how ideas circulate, acting as a good entry into higher levels of both state behavior in general and GEG specifically (Campbell et al., 2014; Corson, Campbell and MacDonald, 2014; Habermas, 1984).

The guiding assumption of 'state learning' is that the more we know about the ensemble of components making out action, and the relations between them, the more we can derive from actions in higher levels. The unlocking of important keys – like that of how to achieve enhanced and efficient Arctic environmental cooperation – is located to the micro-level analysis of state behavior, where social contingent parameter such as trust, awareness, expectations, and notions of self all are imbedded. Following Corson, Campbell and MacDonald (2014:21, 25), to capture the personal in politics is important, since social relationship also affect what will be attributed to those meaning-systems found within political decision-making. Ultimately, the micro level and its focus on interpersonal interactions offers an entry into the domain of learning processes, by bringing to fore how state positions within environmental negotiations – in theory as well as in practice – get impoverished by social interaction. The micro level manifests the state role as 'in the making' and captures and visualizes the process where state representatives in situ are balancing various expectations. Although it is less likely that one negotiation session that is rather isolated in time and space will learn states anything profound regarding (the value) of a particular environmental norm, there is from an IR perspective yet much to learn from a micro-level analysis of these shorter negotiations. Above all, it contributes with an understanding of how learning of environmental norms firstly runs through a social learning on roles, and more precisely, on a contextually anchored role suitability. In that way, the micro level eloquently presents the role as a flexible construction, informed by situated understanding and thus, potentially, ready to change, i.e., to learn something new regarding role purpose and content.

### The Learning Link between 'the State' and 'State Representatives'

To search for explanatory sources to state behavior at the micro level increases the pressure on the linkage between the state and state representatives. At first glance this linkage seems full of disruptions; really, would it at all be possible to investigate *state* learning by references to how individual state *representatives* experience the social setting of which they are forming part? Although states are organized through individuals, a state cannot be reduced to individuals, since states are bigger than the sum of interaction that takes place among them (see List and Spiekermann, 2013; Wendt, 2004). States can only act like *as-if* persons, i.e., on the one hand they are the behavior and discourses of the people who make them up, people who furthermore are driven by (cognitive) motivations of what is meant by being a social human being, but on the other hand, states are *also* adding on norms and structures that form (individual) behavior (Wendt, 2004). Therefore, an agent is never bigger than his or her surrounding, just like how structures have effects that cannot be reduced to agents (ibid., 1999:139). These inbuilt unit-level constrains are also what makes it possible to create an analytical linkage between states and state representatives, where states are illustrated as acting like as-if persons.

In social interaction individuals respond to social stimuli, for instance, by holding beliefs and aiming for belongingness and recognition. States are, to use the words of Medby, "inherently peopled," where alongside state personnel's professional role of carrying the duty to represent the state well, they also bring personal experiences and imaginaries into the state (2018:117, 123). Thus states – being an extension of individuals – are responding to social stimuli, longing for recognition, social belonging, and the feeling of inclusion rather than exclusion (Iser, 2015; Lindemann, 2011). For instance, through membership in an organization, states sometimes can be fostered "to act outside their perceived self-interest, since there is much to gain from being understood as a reasonable actor in international relations" (Faure and Lefevere, 2011:74). Such fostering begins at the micro level. Thus behavior, be it state or individual, is to a certain extent driven by material forces that simply cannot be negotiated away since they are part of human/state nature. Things like self-esteem and social belongingness constitute the material ground from which actors in international relations assess interests: interests, as Wendt explains (1999:130), may be constituted by changeable ideas, but they must nonetheless "ultimately hook on to a material ground."

In situations where international norms are an integral part of international relations, actors have become aware of rules and they have been socialized into identifying themselves as members of international society, i.e., as rule followers (Finnemore and Sikkink, 1998). This is true for states, who are prone to peer pressure from other states, attempting to be perceived as legitimate; honor and self-esteem are universal needs that are fulfilled once respect is gained from those who (is perceived to) matter (Lebow, 2008; Browning, 2015). This is also true for individuals, representing the state, as they do not wish to be associated with a behavior being faulty of unattractive.

Indeed, what makes the connection between state and state representative strong is the assumption that state interaction, like international negotiations, bring forward 'the best' in individuals by offering an arena to "demonstrate their personal attractiveness" where they furthermore could show "their intelligence, their cleverness, their coolness" (March, 1994:215). In their role as state representatives, they proclaim their value as individuals – at least those values that are highly cherished socially (ibid.). Individuals are here conformed to succumb under the 'the state,' because two watchwords that few state officials want to violate are public trust and professional integrity (March and Olsen, 1995:58). Individuals representing states could therefore be assumed to act with their respective state's policy preferences and good reputation in mind, treating it as a norm on professionalism not to be violated. Within this lies a willingness to also play the state role as rehearsed, expected, and deemed appropriate.

## Cases and Methodology

### *Introducing the Arctic Council*

Since the AC was established in 1996 it has, over the years, been further developed in terms of both depth and width and become more institutionalized and encompassing in terms of issues of coverage pertaining to sustainable development and environmental protection. To its character, the AC is intergovernmental which in practice means the organization to favor decision-shaping procedures over decision-making ones; its mandate does not stretch to enforcement or monitoring of policies and regulations (Brigham et al., 2016; Knecht, 2017). However, it bases its work on an inclusive soft-law approach (Koivuorova and Heniämäki, 2006), where also associations of indigenous people – the so called Permanent Participants (PPs) – can give their input on the agenda and have an effect on how Arctic issues are understood and approached (ibid.).[5] They are traditional knowledge holders, represented in the AC by Aleut International Association (AIA), Arctic Athabaskan Council (AAC), Gwich'in Council International (GCI), the Inuit Circumpolar Council (ICC), Russian Association of Indigenous Peoples of the North (RAIPON), and the Saami Council. Although no formal decision-making power is allotted to them, one representative explains their impact:

> I have never seen the Arctic states to move forward with something that PPs would disagree on without dealing with the reasons for those disagreements [...]. In a consensus-based context, PP's have a lot of power, more than people realize.

Consequently, this, in combination with all the exhaustive and extensive scientific material produced by research entities within the AC, called Working Groups (WGs), moves Arctic environmental diplomacy beyond state interests and on to norms and prescriptions of having these norms followed.

This book argues the Arctic Council, as well as state relations therein, to be well-suited to study from a (social) learning perspective; the forum is rather small and personal, often with continuity regarding representation, and it also comprises multiple of norms – often conflicting ones given the earlier mentioned 'Arctic paradox' – which states and state representatives yet are forced to relate and respond to. The outspoken wish of the Arctic states to create strong Arctic governance, furthermore, entails a view of Arctic cooperation as utterly unique and therefore worthy of dedicated state support. A quick presentation of the AC and its institutional structure will help the reader to visualize 'state learning,' i.e., the studying of learning as a process, as a combined sum of interaction at different analytical levels.

The Arctic Council is structured as a hierarchical pyramid, around bodies performing both political and scientific work on Arctic issues (Figure 1.1). At the top of the decision-making chain are the *Ministers* located (since 2011, these ministers often are Ministers of Foreign Affairs). Ministerial meetings take place every second year and is hosted by the country holding the chair.[6] The chairmanship offers the state an opportunity to put a mark on how the AC is developing, through chairmanship programs developed which list thematic areas and issues of choice, i.e., cooperative priorities to the state holding the chairmanship. The Ministerial meeting ends with the signing of a declaration, which reflects the cooperative direction for the coming two years, and, if any, agreements, while chairmanship also rotates.

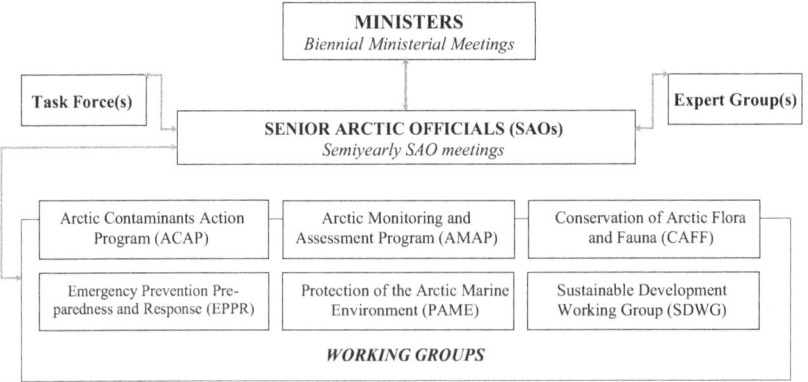

*Figure 1.1* Structure of the Arctic Council.

At the pyramid base are WGs, which perform scientifically based assignments and evaluations of the Arctic region, developing reports and making assessments. As such, they map, investigate, and recommend, extensively contributing to the Arctic knowledge base, providing a background and enhanced understanding for actors operating on a more political mandate. Senior Arctic Officials (SAOs) are national high-representatives, often Arctic Ambassadors, and they are the hub of the Arctic Council: everything that happens in the AC runs through the SAOs. Projects undertaken by the WGs are furthermore presented to SAOs, in a condensed format, who then ask questions or raise concerns. Before each Ministerial meeting, SAOs put together an extensive report based upon all work conducted by the WGs and others, with deliverables and work plans, to the ministers. Lastly, *task forces* are negotiation forums, and as such equipped with a political dimension. They are brought into existence by specific terms of reference, sanctioned by the SAOs and on a mandate that stretches for two years. Task force conclusions are suggestions that in one way or another pertain to how the Arctic cooperation should move forward within specific issue areas, for instance, oil spill prevention and reduction of short-lived climate pollutants. Currently, one *Expert group* is also active, as a direct follow-up on – and administration of – the task force on reductions of black carbon and methane.

### *Introducing the Cases: SAO Meetings and Task Force Negotiations*

This book engages with a question such as *how do states learn of environmental norms* by referring to learning as a social process. Social interaction discloses the strength of not only current environmental norms, but also reveal a variety of other matters that is affecting the development of environmental governance, such as national interests, state reputation, and self-conceptions. To answer the question on how states learn of environmental norms is therefore most suitably done, so this book argues, through a study of how states and state representatives reflect about their role when being subjected to expectations, domestic as well as external ones. Two separate analyses will be carried out, that both operationalizes learning through roles but in different ways. The first one (Chapter 4) concerns an identification – a mapping – of the Arctic states' cooperative roles, as they have been performed in an institution like the Arctic Council. This is a descriptive analysis which applies a text analytical method, relying foremost on SAO meeting minutes.[7] The perpetuation of SAO meetings, twice each year during a period of about twenty years, makes it possible to identify a performative role pattern, rather than describing a role as it was expressed at a specific time and space. The role mapping gives input to subsequent micro-level analyses and discussions of learning (Chapters 5 and 6), but it also gives insight on how social interaction and (perceptions of) others' expectations is pushing roles and behavior in a cooperative direction.

The second study is a dedicated analysis of the consensus-reaching process of two AC negotiations (Chapters 5 and 6) and seeks to reason around learning processes, partly departing from previous chapter and the cooperative roles identified therein. However, it also relies on elite interviews to capture social interaction 'from within,' and its micro-level perspective is believed to offer a more in-depth understanding regarding *how* state actors learn. Two negotiations within politically appointed task forces are investigated: the Task Force on Arctic Marine Oil Pollution Prevention (TFOPP) and the Task Force for Action on Enhanced Reduction of Black Carbon and Methane (TFBCM). Both were established through the Kiruna Ministerial meeting in 2013 and completed by the Iqaluit Ministerial in 2015, and cover issues that when negotiated both were – and still is – of significant importance to the Arctic: short-lived climate pollutants, since these create warming and ice-melting in a region highly adapted to the presence of ice; and oil spill prevention, since marine shipping as well as energy extraction in the offshore Arctic, where a vast majority of resources a relocated, means a substantial risk for oil spills (Hasselström et al., 2017; Kakabadse, 2015; Kutepova et al., 2011). These two negotiation processes *together* serve to enhance understanding on learning processes and should therefore be viewed as complementary cases.

### Interpreting Interview Data

From a role theoretical perspective, states' role conceptions are more multidimensional than derived material facts only, such as territorial size and resource muscles. Instead, a role conception is a combined sum of contextual interpretations, centered on questions such as what type of state actor am I given domestic expectations, what kind of state actor do *I* *think* external actors see me as, and what, *really*, is expected out of me on the international arena? These are all considerations that form a role conception, and thus affect how the role will be performed. As a method, interviews therefore would be a good choice to reveal state interpretations of role conceptions, by giving access to reasoning and reflections of roles as social artifacts. The descriptive text analysis of Chapter 4 would in this context be vital to provide role conceptions with a background and assumed substance.

For this study on learning processes, about thirty interviews have been conducted with participants in TFOPP and TFBCM, including PPs, WGs, and observers, in addition to a handful of interviews with people previously engaged in AC cooperation or with insight thereof. Most of the interviewees were state representatives, with professional titles like advisers and senior advisers, senior consultants, and directors.[8] As such, they fit the definition of elite interviews by being either in close proximity to power and policy-making (Lilleker, 2003), or having particular expertise (Morris, 2009). Through these interviews, learning and role-playing is sought to be understood from within, by "scrutinizing and dissecting social mechanisms, such as meaning-making" (Wacquant, 2002:1470).[9] For instance, how do state actors reason about the process of which they are part of; how do their own state interests, perceptions, and role characteristics fit within the specific process, and how do understanding – of others, the process, and of themselves – create changes in role performance, or ultimately, in beliefs? Such an analysis can thus be described as an endeavor to "make sense of how others make sense of the world" (Jackson, 2008:91).

However, the validity of the interview material such an analysis draws upon lingers on epistemological reflections concerning how these interview accounts came about. Does, for instance, the interview *generate* data, or is the act of interviewing itself *the* data? is it a resource, or a topic (see, for instance, Rapley, 2001)? Similarly, is the interviewer a miner – someone who 'digs out' (reveals and collects) the knowledge hidden within the respondent, or is the interviewer a traveler who sees the interview as a conversation with no predestined direction – as a result constructed by the interaction between the interviewer and interviewee (Kvale, 2009)? This book adheres to a position where knowledge is possible to 'collect' by listening to the interviewee and his or her displayed perspective (i.e., his or her *version* of data). To exemplify, the interview is believed to be capable of providing and generating knowledge on meeting procedures, state preferences, and expectations, as well as the importance being ascribed to the negotiated issue. However, as the example here tries to illustrate, to 'collect knowledge' is not without difficulties:

---

**BOX 1.1 The need of interview reflexivity**

*In the Far North, a phone interview with Mr M had been scheduled. Following the interview guide, questions asked did not differ much in substance from questions that previously had been asked to – and answered by – other interviewees. However, to Mr M they were troublesome. Troublesome to a point where he halfway into the interview said: "I do not know if this is going to be worth your time. I am going to be honest with you: my emotional sensing is not that high and I am not a very perceptive man on reading people." This interview left a bitter taste hard to get rid of, not just because what he said but because how he said it. Mr M was probably slightly irritated with me for asking questions he could not answer or had an interest in. However, his comment was also said with a certain amount of resignation, that he couldn't provide me with those answers he thought I was after. Thus, not just had I unintentionally stirred up awkward feelings in the interviewee, his reaction made it apparent that an interview is more than a data-generating source on a specific topic. It was an insight that complicated my previous understanding of interviews as for sure subjective but still factual. Now, however, Mr M's comment had turned the interview personal. In*

> *two short sentences, he demonstrated that what I wanted for an interview was without reach if not simultaneously attuned toward the way the data came about.*

If the interview mentioned had been all about revealing information possessed by Mr M – and where his words had been taken to ensure real experiences – not much room had been given to think about the impact of the researcher on these told experiences. This is critically referred to as the 'vessel of answer' approach, where one fails to view the interviewee as an active subject that in an *interactive* setting actually partakes in producing knowledge (Holstein and Gubrium, 1995). To return to Mr M: Mr M quite evidently felt reluctant to answer questions about the social dimension of international negotiations – but he still repeatedly *tried* to. What does this mean for the interview result?; was he pushed in a direction of saying what he thought was expected of him? Did my choice of words and questions create an interview knowledge that was constructed to such an extent that it had less to do with the interviewee's experiences and more to do with our interaction? When interpreting the interview data, these were the questions to consider.

"The point is to try to grasp the relationship between the interview as social interaction and as vessel of topic," Warren explains (2012:130), referring to the difficulty of knowing if the respondent's answer is suitable to use as a resource, as data. Such 'grasping' should be guided by *reflexivity,* which allows for the subject matter to be interpreted from different angles, challenging prevailing interpretations, and spur imaginative efforts in order to extend the thinking of specific phenomena (Alvesson, 2011:106). Reflexivity would also be important in order to correctly interpret which subject is given voice in the interview. Mr M, for instance, is giving the interview in the capacity of him being a state representative. In the previous interview quote, however, he refers to *himself* as the subject rather than the role of being a state representative. To separate roles – capacities in which people speak – could be difficult during an interview. It is, however, important to be attentive to which of the subjects that is being talking since this could have an effect on how the data should be interpreted.

Two other examples will be given in the following, where a reflexive approach on social interaction is argued to be important. *Example 1:* When sharing views on the presence of energy interests within the task force on reductions of emissions from black carbon and methane, one participant in a leading role expressed the following:

> Well, it has not been as noticeable as I thought it would be. But that is because we have concentrated on this issue [reducing black carbon and methane], we have not discussed a ban on Arctic oil prospecting or production. Then, of course, it had been much clearer lines of division between those who want it and those who do not (laughing subtly).

Although a first read might suggest that the interviewee did not consider energy politics to have impacted upon negotiations, a second read slightly adds nuances. This is because the interviewee starts off by telling us that he started off from *somewhere* in his expectations of knowing how energy politics would have an impact. Compared to this *starting point,* energy politics was not as noticeable as would have been expected. Perhaps of little surprise, there is a background from which all expectations depart. These expectations could be linked to prior experiences of 'energy politics' noticed within the national/international environmental processes, or they could more loosely be linked to general expectations of how energy interests operate. Nonetheless, to really grasp what the interviewee describes would require a more thorough knowledge of where his expectations start out from. "Don't argue with the members" is an advice by Gubrium and Holstein (2012), pointing toward the importance of context as sense-maker of meaning. One should not challenge the accounts given by the interviewee by replacing it with one's own meaning, but rather try to track where it contextually stems from. Following Alvesson (2011:113–114), to apply a reflective approach here means to imagine the interviewee as, on the one hand, a truth-teller (regarding lived experiences) and, on the other hand, as a political actor that is being constrained by certain structures (regarding what he sees). What is in between is then possible to think about and reflect upon in different ways; to stretch imagination as well as to eradicate misunderstandings and interpretative mistakes. Because elite interviews, where people often have an agenda on their own, are knowledgeable and accustomed talkers, it becomes possibly even more important to be aware of what 'is in between' (see Mikecz, 2012)

The *second example* has to do with the social dimension of interview as method. In many ways, it is similar to conversations. It therefore triggers ordinary social behavior; people (normally) smile when smiled at, are more right to the point when the conversational partner seems in a hurry, and they lower voices in mirror. Then, of course, words used are also copied. The following interview transcript is an attempt to illustrate the latter. Here, the task force interviewee had just talked about different state behaviors, or roles, and I, the interviewer, was interested in finding out more about how this behavior was, or became, cooperative.

INTERVIEWER: Would you say there are aspects of adaptation to each other, perhaps even learning?
INTERVIEWEE: [...] We entered this process with the resolution of achieving a binding quantitative goal, but gradually we understood it would never pass the U.S. Congress. It needed to be formulated differently. One learns from each other during the course of action and realizes that it needs to be something that suits us all, or it will end up being nothing.
INTERVIEWER: You said "learn from each other," can you elaborate on what you mean? Would you, for example, say that learning such as thinking in new ways has been present?

INTERVIEWEE: Well, I have not really thought about it like that. It has been a successive process for a year and a half...But yes, for sure, I believe one has a different thinking at the end of the process than in the beginning, yes definitively, I believe that.

The second part of this transcript is close to the interviewer asking leading questions by telling the interviewee which learning 'to count.' Learning, the interviewer says, equalizes thinking in new ways. Thereby is interaction potentially leading astray from the interviewee's real experiences and perceptions thereof, rendering the data questionable. The first part, however, feels more as originating from the interviewee, but the question is still to what degree. The interviewee mentions learning, that national delegates 'learn from each other' in international negotiations, something liked to be conceptualized further. However, to what extent is the interviewee's account of learning really 'worth' something, since initially the word was coming from the interviewer? How much more value would instead be added in the case if learning firstly appeared in the interviewee's vocabulary, and then picked up by the interviewer? In this particular transcript, it is believed that social interaction has rendered the concept of learning quite empty. But by applying a reflexive approach, while being attentive to and aware of where mistakes have been carried out, interviews are a rich source to gain knowledge about events and experiences that one cannot acquire on one's own.

## Chapter Outline

This *first chapter* has served to set the frames for this book – its topic, ambitions, and theoretical premises. By having introduced the reader to the main concepts and assumptions that will guide subsequent discussions on Arctic state behavior and learning, it should now be obvious – on a general level – as to what to expect in a book where expectations indeed are held to matter. In Chapter 2, the theoretical underpinnings of roles will be discussed more in depth, especially in relation to how the role concept contains manifestations of learning. Arguing, understanding, expectations, and norms constitute the forming qualities for the way roles and learning are expressed. The chapter presents an analytical model for how a phenomenon like learning that often occurs within actors, in their belief system, is suggested to be studied. Chapter 3, then, is an empirical chapter on Arctic environmental cooperation and the so called 'Arctic paradox,' which stipulates that any attempt of strengthening environmental protection by, for instance, mitigating Arctic warming, firstly needs to overcome those economic opportunities that are laid bare through warmer temperatures. The chapter centers on the question of what kind of impact this (presumable) dichotomy has on Arctic cooperation.

In Chapter 4, focus is on the performance pattern of each Arctic Council member state. Here state roles will be mapped according to how the role is conceived upon nationally, and how it has been expressed in Senior Arctic

Officials' meetings. This section provides for a descriptive analysis, drawing on foremost minutes from meetings but also on national documents such as Arctic strategies. In Chapters 5 and 6, the reader gets transferred to the micro level through an investigation of the processes where negotiating state representatives attempt to reach consensus: on oil pollution prevention and reduction of short-lived climate pollutants, respectively. Rather than comparing these different negotiations processes, these chapters serve to concretize the learning process by investigating what social interaction does for (a contextually anchored) understanding, and how social interaction may add nuances or changes to the roles previously mapped in Chapter 4.

The last part of this book is a summary devoted to state learning on environmental norms. In Chapter 7, the analytical model that firstly was presented in Chapter 2 will be assessed and discussed – also in relation to the empirical learning of the Arctic environment, given what then has been known from previous chapters. The final *epilogue* takes somewhat of a retrospective approach, glancing back on what, if anything, the Arctic states really seem to have learnt regarding Arctic environmental protection. Above all, it looks beyond the horizon, tentatively suggesting the Arctic region to be headed in a direction of continuous cooperation and where the stakes are raising for anyone that violates the (rising) norm of environmental responsibility.

## Notes

1  In connection to black carbon particles triggering Arctic warming, and supported by IPCC's findings, Finland's president, Sauli Niinistö, put it in the following way when speaking at a press conference together with President Trump: "And the problem is not only Arctic; if we lose the Arctic, we lose the globe" (Niinistö, 2017).
2  Every second year, in Springtime, the Arctic states meet at Ministerial meetings, simultaneously rotating chairmanship in two years' cycles. At the meetings, it is customary for ministers from each state to give a speech, and a recurrent theme in these speeches is to describe Arctic cooperation as good, peaceful, and most important for the environment (see, for instance, member statements, AC Ministerial meeting, 2015; 2017).
3  Held in the early 1970s, the Stockholm Conference is often referred to as a starting point for an issue like global environmental responsibility. The transboundary character of environmental problems was emphasized, as well as the subsequent need for international cooperation and regulation. Nevertheless, the conference *also* acknowledged each state's sovereign right to exploit its own natural resources, as long as this would occur "within the limits of not causing damage beyond its national borders" (Stockholm Declaration, 1972).
4  For a similar conclusion on states in relation to global governance, see Rottem (2017), who discusses the way science is travelling into norms and regulatory frameworks
5  Apart from 6 PPs, and of course the 8 Arctic states, the AC hosts a number of observers (38 in total as of January 2020), both states and organizations, where the interest to become an observer has steadily increased over the years.

6 The order of chairmanship has been the following: Canada 1996–1998, U.S.1998–2000, Finland 2000–2002, Iceland 2002–2004, Russia 2004–2006, Norway 2006–2009, Denmark 2009–2011, Sweden 2011–2013, Canada 2013–2015, U.S. 2015–2017, Finland 2017–2019, and Iceland 2019–.

7 The collected material has not before been investigated in any vigorously extent. Prior this, a questionnaire-survey has investigated the former Artic Council members and participants and their views on the Arctic Council being an effective governance system (Kankaanpää and Young, 2012); a recent dataset study has provided an analysis on stakeholders' participation in AC meetings (Knecht, 2017); and meeting minutes from chosen time periods have been investigated in relation to Russia's cooperative Arctic commitments (Chater, 2016).

8 Interviews were prioritized with representatives with high meeting attendance. In addition, the aim was to the extent possible have interviewees distributed equally between the states, and preferably with HoD from each state. More information on the way interviews were conducted and distributed between state members is provided for in respective Chapters 5 and 6.

9 Interviews were semi-structured, and the interview guide centered around four themes: key message from state X in relation to TFOPP/TFBCM; how the task force process was experienced; which (implicit) role state X perceived itself to be given and/or perform; and 'lessons learned.' The focus in interviews was thus on interviewees' reflections on what it meant to represent a specific state in a specific group.

## References

AC Ministerial Statement by Finland. (2017). *Statement by Mr. Timo Soini, Minister for Foreign Affairs of Finland, Arctic Council 10th Ministerial meeting, Fairbanks, Alaska, 11 May 2017.* Accessed via oaarchive.arctic-council.org.

Aggestam, L. (2006). "Role Theory and European Foreign Policy: A Framework of Analysis," in Elgström, O. and Smith, M (eds.). *The European Union's Roles in International Politics. Concepts and Analysis.* London and New York: Routledge.

Albert, M. and Vasilache, A. (2017). "Governmentality of the Arctic as an International Region," *Cooperation and Conflict,* Vol. 53:1, pp. 1–20.

Alvesson, M. (2011). *Interpreting Interviews.* London: Sage.

Arctic Council. (2004). *ACIA – Impacts of a Warming Arctic. Arctic Climate Impact Assessment (ACIA) Overview Report.* Cambridge: Cambridge University Press, pp. 1–140.

Arctic Monitoring and Assessment Programme (AMAP). (2017). *Snow, Water, Ice and Permafrost in the Arctic. Summary for Policy-Makers.* Oslo, Norway.

Axworthy, T. S., Koivurova, T. and Hasanat, W. (eds.). (2012). *The Arctic Council: Its Place in the Future of Arctic Governance.* Collection of papers originally presented during a conference organized by Munk-Gordon Arctic Security Program, University of Lapland.

Barry, T., Davíðsdóttir, B., Einarsson, N. and Young, O. R. (2020). "The Arctic Council and Agent of Change?" *Global Environmental Change,* Vol. 63:July, pp. 1–10.

Borgerson, S. G. (2008). "Arctic Meltdown: The Economic and Security Implications of Global Warming," *Foreign Affairs,* Vol. 87:2, pp. 63–77.

Brigham, L., Exner-Pirot, H., Heininen, L. and Plouffe, J. (2016). "Introduction – The Arctic Council: Twenty Years of Policy-Shaping," in Heininen, L., Exner-Pirot, H. and Plouffe, J. (eds.). *Arctic Yearbook 2016. The Arctic Council: Twenty Years of Regional Cooperation and Policy-Shaping.* Iceland: Northern Research Forum.

Browning, C. S. (2015). "Nation Branding, National Self-Esteem and the Constitution of Subjectivity in Late Modernity," *Foreign Policy Analysis,* Vol. 11:2, pp. 195–214.

Burke, D. C. (2019). *Diplomacy and the Arctic Council.* Canada: McGill-Queen's University Press.

Campbell, L. M., Corson, C., Gray, N. J., MacDonald K. I. and Brosius, P. J. (2014). "Studying Global Environmental Meetings to Understand Global Environmental Governance: Collaborative Event Ethnography at the Tenth Conference of the Parties to the Convention on Biological Diversity," *Global Environmental Politics,* Vol. 14:3, pp. 1–20.

Chater, A. (2016). "Explaining Russia's Relationship with the Arctic Council," *International Organizations Research Journal,* Vol. 11:4, pp. 41–54.

Chuffart, R. (2019). "Is the Arctic Council a Paper Polar Bear?" *High North News,* November 29, 2019.

Clapp, J. and Dauvergne, P. (2011). *Paths to a Green World. The Political Economy of the Global Environment,* 2nd ed. Cambridge, MA: MIT Press.

Clark, W. C., Mitchell, R. B. and Cash, D. W. (2006). "Evaluating the Influence of Global Environmental Assessments," in Mitchell, R. B., William C. C., Cash, D. W. and Dickson, N. M. (eds.). *Global Environmental Assessments: Information and Influence.* Cambridge, MA: MIT Press.

Corson, C., Campbell, L. M. and MacDonald, K. I. (2014). "Capturing the Personal in Politics: Ethnographies of Global Environmental Governance," *Global Environmental Politics,* Vol. 14:3, pp. 21–40.

Dodds, K. J. (2013). "Environment, Resources, and Sovereignty in the Arctic Region: The Arctic Council as Regional Body," *Georgetown Journal of International Affairs,* Vol. 14:2, pp. 29–38.

Dubois, M-A. and Tesar, C. (2014). "Making It Stick – A New Approach to Implementing Arctic Council Decisions and Recommendations," in Heininen, L., Exner-Pirot, H. and Plouffe, J. (eds.). *Arctic Yearbook 2014. Human Capital in the North.* Iceland: Northern Research Forum.

Falkner, R. (2012). "Global Environmentalism and the Greening of International Society," *International Affairs,* Vol. 88:3, pp. 503–522.

Farkas, A. (1998). *State Learning and International Change.* Ann Arbor, MI: University of Michigan Press.

Faure, M. G. and Lefevere, J. (2011). "Compliance with Global Environmental Policy," in Axelrod, R. S., VanDeveer, S. D. and Downie, D. L. (eds.). *The Global Environment. Institutions, Law and Policy,* 3rd ed. Washington, DC: CQ Press.

Finnemore, M. and Sikkink, K. (1998). "International Norm Dynamic and Political Change," *International Organization,* Vol. 52:4, pp. 887–917.

Franchini, M., Viola, E. and Barros-Platiau, A. F. (2017). "The Challenges of the Anthropocene: From Environmental Politics to Global Governance," *Ambiente & Sociedade,* Vol. 20:3, pp. 177–199.

Gad, U. P. and Strandsbjerg, J. (eds.). (2019). *The Politics of Sustainability in the Arctic. Reconfiguring Identity, Space, and Time.* London and New York: Routledge.

Gorbachev, M. (1987). *Mikhail Gorbachev's Speech in Murmansk at the Ceremonial Meeting on the Occasion of the Presentation of the Order of Lenin and the Gold Star to the City of Murmansk.* October 1, 1987.

Gubrium, J. F. and Holstein, J. A. (2012). "Don't Argue with the Members," *American Sociologist,* Vol. 43:1, pp. 85–98.

Habermas, J. (1984). *The Theory of Communicative Action. Reason and the Rationalizaton of Society,* Vol. 1. Cambridge: Polity Press.

Harnisch, S. (2015). "Role Theory and the Study of Chinese Foreign Policy," in Harnisch, S., Bersick, S. and Gottwald, J-C. (eds.). *Chinas International Roles. Challenging or Supporting International Order?* Florence: Taylor and Francis.

——— (2012). "Conceptualizing in the Minefield: Role Theory and Foreign Policy Learning," *Foreign Policy Analysis,* Vol. 8:1, pp. 47–69.

——— (2011). "Role Theory: Operationalization of Key Concepts," in Harnisch, S., Frank, C. and Maull, H. W. (eds.). *Role Theory in International Relations. Approaches and analyses.* London and New York: Routledge.

Harnisch, S., Frank, C. and Maull, H. W. (2011). "Conclusion: Role Theory, Role Change, and the International Social Order," in Harnisch, S., Frank, C. and Maull, H. W. (eds.). *Role Theory in International Relations. Approaches and Analyses.* London and New York: Routledge.

Hasselström, L., Håkansson, C., Noring, M., Soutukorva, Å. and Khaleeva, Y. (2017). "Costs and Benefits Associated with Marine Oil Spill Prevention in Northern Norway," *The Polar Journal,* Vol. 7:1, pp. 165–180.

Holstein, J. A. and Gubrium, J. F. (1995). *The Active Interview.* Thousands Oak, CA; London: Sage.

Humrich, C. (2013). "Fragmented International Governance of Arctic Offshore Oil: Governance Challenges and Institutional Improvement," *Global Environmental Politics,* Vol. 13:3, pp. 79–99.

Intergovernmental Panel on Climate Change (IPCC). (2018). *Summary for Policymakers of IPCC Special Report on Global Warming of 1.5°C Approved by Governments.* IPCC Press release, 2018/24/PR, October 8, 2018, www.ipcc.ch/site/assets/uploads/2018/11/pr_181008_P48_spm_en.pdf.

Interviews. (2014–2015). Interviews between December 2014 and December 2015 with state representatives and other representatives active in Arctic Council cooperation and negotiation.

Iser, M. (2015). "Recognition between States? Moving beyond Identity Politics," in Daase, C., Fehl, C., Geis A. and Kolliarakis, G. (eds.). *Recognition in International Relations. Rethinking a Political Concept in a Global Context.* Houndmills, Basingstoke: Palgrave Macmillan.

Jackson, P. T. (2008). "Can Ethnographic Techniques Tell Us Distinctive Things about World Politics?" *International Political Sociology,* Vol. 2:1, pp. 91–93.

Kakabadse, Y. (2015). "Frontier Mentality Has No Place in the Arctic," *Harvard International Review,* Vol. 36:3, pp. 55–59.

Kankaanpää, P. and Young, O. R. (2012). "The Effectiveness of the Arctic Council," *Polar Research,* Vol. 31, pp. 1–14.

Knecht, S. (2017). "The Politics of Arctic International Cooperation: Introducing a Dataset on Stakeholder Participation in Arctic Council Meetings, 1998–2015," *Cooperation and Conflict,* Vol. 52:2, pp. 203–223.

Knopf, J. W. (2003). "The Importance of International Learning," *Review of International Studies,* Vol. 29:2, pp. 185–207.

Koivurova, T. and Heinämäki, L. (2006). "The Participation of Indigenous Peoples in International Norm-Making in the Arctic," *Polar Record,* Vol. 42:221, pp. 101–109.

Koivurova, T. and Molenaar, E. J. (2010). *International Governance and Regulation of the Marine Arctic.* Three reports prepared for the WWF's International Arctic Programme. Oslo: WWF International Arctic Programme.

Kutepova, E. A., Knizhnikov, A. Yu. and Kochi, K. V. (2011) *Associated Gas Utilization in Russia: Issues and Prospects.* Annual Report, Issue 4. Moscow: WWF Russia-KPMG.

Kvale, S. (2009). *Interviews. Learning the Craft of Qualitative Research Interviewing,* 2nd ed. Los Angeles: Sage.

Langhelle, O., Blindheim, B-T. and Øygarden, O. (2008). "Framing Oil and Gas in the Arctic from a Sustainability Perspective," in Mikkelsen, A. and Langhelle, O. (eds.). *Arctic Oil and Gas – Sustainability at Risk?* London and New York: Routledge.

Lebow, R. N. (2008). *A Cultural Theory of International Relations.* Cambridge: Cambridge University Press.

Lilleker, D. G. (2003). "Interviewing the Political Elite: Navigating a Potential Minefield," *Politics,* Vol. 23:3, pp. 207–214.

Lindemann, T. (2011). "Peace through Recognition: An Interactionist Interpretation of International Crises," *International Political Sociology,* Vol. 5:1, pp. 68–86.

List, C. and Spiekermann, K. (2013). "Methodological Individualism and Holism in Political Science: A Reconciliation," *The American Political Science Review,* Vol. 107:4, pp. 1–15.

March, J. G. (1994). *A Primer on Decision-Making: How Decisions Happen.* New York: Free Press.

March, J. G. and Olsen, J. P. (1995). *Democratic Governance.* New York: Free Press.

Medby, I. A. (2018). "Arcticulating State Identity: 'Peopling' the Arctic State," *Political Geography,* Vol. 62, pp. 116–125.

Mikecz, R. (2012). "Interviewing Elites: Addressing Methodological Issues," *Qualitative Inquiry,* Vol. 18:6, pp. 482–493.

Morris, Z. S. (2009). "The Truth about Interviewing Elites," *Politics,* Vol. 29:2, pp. 209–217.

Niinistö, S. (2017). *Remarks by President Trump and President Niinistö of Finland in Joint Press Conference.* August 28, 2017. The White House, Office of the Press Secretary.

Nilsson, A. E. (2012). "Knowing the Arctic: The Arctic Council as a Cognitive Forerunner," in Axworthy, T. S., Koivurova, T. and Hasanat, W. (eds.). *The Arctic Council: Its Place in the Future of Arctic Governance.* Collection of papers originally presented during a conference organized by Munk-Gordon Arctic Security Program, University of Lapland.

——— (2007). "A Changing Arctic Climate. Science and Policy in the Arctic Climate Impact Assessment." Dissertation. Linköping Studies in Art and Science No 386. Department of Water and Environmental Studies, Linköping University. Linköping: UniTryck.

Nilsson, A. E. and Koivurova, T. (2016). "Transformational Change and Regime Shifts in the Circumpolar Arctic," *Arctic Review of Law and Politics,* Vol. 7:2, pp. 179–195.

Nord, D. C. (2016a). *The Arctic Council: Governance within the Far North.* New York: Routledge.

———— (2016b). *The Changing Arctic. Consensus Building and Governance within the Arctic Council.* New York: Palgrave Macmillan US.

Onuf, N. (1995) "Levels," *European Journal of International Relations,* Vol. 1: 35, pp. 35–58.

*Ottawa Declaration.* (1996). *Declaration on the Establishment of the Arctic Council.* Signed by the Arctic States on September 19, 1996.

Pattberg, P. (2012). "Transnational Environmental Regimes," in Biermann, F. and Pattberg, P. (eds.). *Global Environmental Governance Reconsidered.* Cambridge, MA: MIT Press.

Pedersen, T. (2012). "Debates over the Role of the Arctic Council," *Ocean Development & International Law,* Vol. 43:2, pp. 146–156.

Posner, E. (2007). "The New Race for the Arctic." *The Wall Street Journal,* August 3, 2007.

Poto, M. P. and Fornabaio, L. (2017). "Participation as the Essence of Good Governance: Some General Reflections and a Case Study on the Arctic Council," *Arctic Review on Law and Politics,* Vol. 8, pp. 139–159.

Rapley, T. J. (2001). "The Art(fulness) of Open-Ended Interviewing: Some Considerations on Analyzing Interviews," *Qualitative Research,* Vol. 1:3, pp. 303–324.

Rottem, S. V. (2017). "The Use of Arctic Science: POPs, the Stockholm Convention and Norway," *Arctic Review on Law and Politics,* Vol. 8, pp. 246–269.

SAO Meeting Minutes. (2016). *Arctic Councils Senior Arctic Officials Meeting Fairbanks, Alaska, U.S.A.* March 16–17, 2016. Report, pp. 1–27. Accessed via oaarchive.arctic-council.org.

Scholte, J. A., Tallberg, J. and Verhaegen, S. (2019). *Elite Attitudes toward Global Governance. A Report of Summary Findings from the LegGov Elite Survey.* Legitimacy in Global Governance Programme. University of Gothenburg and Stockholm University.

Sörlin, S. (2014). "Rättvisa i centrum på miljöpolitikens nya spelplan," in Nordin, T. (ed.). *Miljöpolitikens Spelplan.* Miljöforskningsberedningen, Report October 2014. Stockholm: Fritzes.

Speth, J. G. and Haas, P. M. (2006). *Global Environmental Governance. Foundations of Contemporary Environmental Studies.* Washington: Island Press.

Stokke, O. S. (2014). "International Environmental Governance and Arctic Security," in Tamnes, R. and Offerdal, K. (eds.). *Geopolitics and Security in the Arctic. Regional Dynamics in a Global World.* London and New York: Routledge.

———— (2013). "Regime Interplay in Arctic Shipping Governance: Explaining Regional Niche Selection," *International Environmental Agreements: Politics, Law and Economics,* Vol. 13:1, pp. 65–85.

Thies, C. G. (2010). "Role Theory and Foreign Policy," in Denemark, R. A. (ed.). *The International Studies Encyclopedia.* Blackwell Reference Online.

Wacquant, L. (2002). "Scrutinizing the Street: Poverty, Morality, and the Pitfalls of Urban Ethnography," *American Journal of Sociology,* Vol. 107:6, pp. 1468–1532.

Walker, S. G. (1992). "Symbolic Interactionism and International Politics: Role Theory's Contribution to International Organization," in Cottam, M. and Chih, C-Y (eds.). *Contending Dramas: A Cognitive Approach to Post-war International Processes.* New York: Praeger.

Warren, C. A. B. (2012). "Interviewing as Social Interaction," in Gubrium, J. F., Holstein, J. A., Marvasti, A. B. and McKinney, K. D. (eds.). *The Sage Handbook of Interview Research: The Complexity of the Craft.* Thousand Oaks, CA: Sage.

Wendt, A. (2004). "The State as Person in International Theory," *Review of International Studies,* Vol. 30:2, pp. 289–316.

———— (1999). *Social Theory of International Relations.* Cambridge: Cambridge University Press.

Wilson, P. (2016). "Society, Steward or Security Actor? Three Visions of the Arctic Council," *Cooperation and Conflict,* Vol. 51:1, pp. 55–74.

Wood-Donnelly, C. (2016). "From Whale to Crude Oil: Lessons from the North American Arctic," *Energy Research & Social Science,* Vol. 16, June, pp. 132–140.

Wormbs, N. and Sörlin, S. (2013). "Assessing Arctic Futures. Voices, Resources, Governance," in *Mistra Arctic Futures in a Global Context.* Final Report 2011–2013.

Young, O. R. (2016). "The Shifting Landscape of Arctic Politics: Implication for International Cooperation," *The Polar Journal,* Vol. 6:2, pp. 209–223.

———— (2009). "The Arctic in Play: Governance in a Time of Rapid Change," *The International Journal of Marine and Coastal Law,* Vol. 24:2, pp. 423–442.

———— (2008). "The Architecture of Global Environmental Governance: Bringing Science to Bear on Policy," *Global Environmental Politics,* Vol. 8:1, pp. 14–32.

Zelli, F., Gupta, A. and van Asselt, H. (2012). "Horizontal Institutional Interlinkages," in Biermann, F. and Pattberg, P. (eds.). *Global Environmental Governance Reconsidered.* Cambridge, MA: MIT Press.

# 2 The Studying of Learning through Roles

## Introduction

Whenever a word such as learning is used, two somewhat different interest areas – broadly speaking – could be impinged: learning as an empirical outcome; or learning as a social undertaking. Empirically, it could be established that the world has not learned enough, given the continuation of environmental pollution and degradation found in the wake of states, businesses and people alike too often exploiting the nature in an unsustainable manner. This manner goes as well for the Arctic: although Arctic environmental governance has been strengthened over the years there is – from a strictly environmental learning perspective – a need for enhanced governance considering that economic interests and development continue to overtly triumph environmental concerns. It is therefore of importance to enhance understanding of *how* learning occurs, to also gain insight on how it can be increased: indeed, the idea of studying learning could be pinned down to how 'undesirable' behavior – like, for instance, excessive deforestation or market-driven tear and wear consumption – could be avoided while sustainable and 'good behavior' promoted. What is it then that keeps international actors from reaching strong environmental agreements? That in some respects restrains them from taking on the recommendations from environmental science, while accepting others? The analytical focus of this chapter and book centers on learning as a social undertaking; it is here, when learning is approached as a social process and as 'in-the-making' that insight can be provided on how actors get to learn in the first place.

Although there are several concepts within IR that address learning – e.g., socialization, integration, managing the commons, and epistemic communities – learning often takes an implicit form, keeping the process of learning rather in the dark. Therefore, the aim of this chapter is to turn something diffuse and often unobservable into observable matters. It is an attempt to engage with what has been described as a major and tricky thing with learning: to reveal its existence while moving from theory to operationalization (see Checkel, 2003). Calls have previously been raised on greater conceptualization and methodological stringency regarding the learning concept, to enhance analytical transparency (cf. Checkel, 2001;

Harnisch, 2012; Knopf, 2003). Because how can we be assured that the behavior we are witnessing really is a result of learning, and not just a coincidental behavioral change? And what about such learning that cannot easily be traced to definite moments in time and space? Really, to learn does not even need to involve a skill but could just be a thought.

This book and chapter offer an entry into theorizing learning through role theory and symbolic interactionism. The role concept captures how actors assess and reflect on their behavior in a context informed by diverse (social) norms. This theoretical premise does not necessarily contradict, for instance, a rational choice perspective on state cooperation and learning, where the emergence of regimes are linked to states being rational actors; while keeping their individual preferences in mind, states foresee collaboration as the optimal outcome to circumvent competition in the international system. From this perspective, learning equalizes states adapting their strategies to the interactive context. However, as envisaged by this book, learning is also about cognitive processes that change beliefs, something R. Keohane – who is closely associated with a rational theory like neoliberal institutionalism – acknowledged; in his seminal book *After Hegemony*, he called for an examination of what the premise of self-interest means, given that also supposedly objective state interests do change. Referring to the thoughts of E. Haas, he suggested regime change to ultimately be the result of a change in how people think about their interests, i.e., a result of learning (1984:128; see Haas, 1990). Thus, from a rational choice perspective as well, learning is a refined concept that encompasses and acknowledges actors' cognition in multiple ways. But, as conceded by Keohane, no book can do it all (ibid.), but needs to choose where to locate focus. This goes as well for theories.

In fact, since before it has been suggested that theories like neoliberal institutionalism and the constructivist approach have rather similar analytical qualities, merely different interest areas in terms of explaining short-term cooperative behavior versus understandings of long-term change (Sterling-Folker, 2000). This book too, by using 'the role' as a tool for understanding the learning process, has its analytical focus shifted from *what* (specificities) actors learn in the moment to how they (reciprocally) learn. Its main interest is not to categorize and label whether a specific learning is a result of either strategy or changed beliefs – but rather to increase understanding on the process of learning; how does social interaction shape perceptions and behavior, how does norms 'travel' in social interaction, and how does learning connect to change in international relations? These are questions all suitably addressed by role theory. Not least does it uncover an opportunity to have the abstract dimensions of learning theoretically traced and anchored.

Similar to Aggestam (2006), who has argued that "dynamism and process" needs to be brought into the analysis of how roles are learned in interaction processes such as negotiations (ibid.:16), the remaining part of this chapter seeks to explain and further conceptualize the relationship between roles and learning; what are these two concepts about and how are they connected? Exploring this, the chapter applies a micro perspective on international relations, calling for an investigation of the belief system, by drawing on (1) social psychological underpinnings of socialization and (2) the sociological claim of a particular form of communicative rationality to inform the understanding of actors. The chapter is summed up through a theoretically deductive 'learning triangle,' to deliver an analytical model on the (social) process of state learning.

## Roles: A Tracking Tool for Learning

That states and other interacting actors are playing roles – be it the role of a leader, mediator, foot-dragger, or all or none of these – is, of course, metaphorically borrowed from theater (Thies, 2010). In theater, actors follow a script which pinpoints the 'rules' for the assembly to play by. As such, the script entails storylines that to a certain extent decode how roles should be performed in an appropriate way (Hajer, 2005), so that story and end concur. Now, imagine there would be no script to follow when interacting socially – neither in international cooperation nor in negotiations, would states then always be aware of their cooperative interest? This book is inclined to answer that question with a no. Or, as Martha Finnemore has put it, "states do not always know what they want" (Finnemore, 1996:128). Instead 'wants' are something highly contextual, added with meaning as interaction proceeds. Roles thus represent excellent tools of revealing wants; different roles want different things, where international norms add complexity and where group dynamics give further input. On some occasions the collision between these appointed 'wants' and more traditional national interests and perceptions results in too much of a mismatch, pushing states to review the role they play. At times like these, acting as a "social theory of international politics" (Harnisch, 2012:48), role theory sheds light on how change comes about. A close look at roles, and changes therein, could therefore reveal learning.

Role theory entered the field of political science via foreign policy analysis but has been increasingly applied within IR. With the former's actor-centered approach (bottom-up) and the latter's focus on agent-(international) structure (often top-down), these two perspectives are considered to be a good synthesis through which roles in international politics could be understood (Harnisch, Frank and Maull, 2011; Thies and Breuning, 2012). The breakthrough of role theory in political science dates quite far back in time, through Kalevi Holsti's seminal article (1970) on national role conceptions. Holsti recognized several things of importance when it came to foreign policy behavior, for instance, that actors (state leaders, in Holsti's approach) in diplomatic settings were enacting several different roles, that there furthermore were behavioral variation amongst these roles, and that structure mattered to encourage individual behavior "to fit the expectations of others" (Holsti, 1970:235–236). However, he did not really take the *interaction effect* between states into consideration when studying roles: instead, roles originally came from domestic conceptions, which then were adjusted to those external expectations imbedded in the international structure of laws and values. As a response to Holsti's seminal article, role theory was rather quickly argued to have more to offer to studies of international relations, not least when it came to role conflicts and the impact of others (Backman, 1970). Really, the core aspect of what roles can contribute is the bringing in of "others," thus widening the traditional view on state behavior within IR as a result sprung (only) national interests and ideas.

Within this series on role theory and international relations, role theory has fruitfully been applied to a variety of cases spread over different levels of analyses, while being subject for continuous theoretical exploration and development (see, for instance, Cantir and Kaarbo, 2016; de Sá Guimarães, 2020; Walker, 2013). Role theory has, amongst others, been applied to increase understanding on how emerging states become full member of the international system, leaning on international social pressure, i.e., state socialization (Thies, 2014). It has been used to better understand the dynamics of conflicts, like past and present enemy relationships between states (Malici and Walker, 2016), and to understand regional transformation in the aftermath of the Arab Uprising, departing from national role conceptions (Akbaba and Özdamar, 2019). A role theoretical approach has furthermore been applied when examining China's international role (Harnisch, Bersick and Gottwald, 2015), South Korea's international state identity (Flamm, 2019), and state decisions to sign/ratify the Kyoto protocol (Below, 2015). This book shares the ambition of many of the ones above to learn more of the potential for progress in international relations by applying a role theoretical perspective.

### The Role Concept and Its Different Components

An analysis of roles engages with interaction carried out between agent and structure (Breuning 2011; Thies and Breuning, 2012), and by connecting self and society, identity (as in ego) and action (McCourt, 2012). Actors could be defined and approached as "social self-in-situation," where actors are defined through Self (ego), alter (Other), and environment (Situation) (Walker, 2013:25). As such, the role concept moves beyond actors as isolated entities, and instead conduct a trisected analysis of behavior. Figuratively, a role would be a *triptych story of expectations*.

When roles are studied, it is rendered important to firstly investigate the domestic context from where roles – and ideas of roles – initially emerge (Cantir and Kaarbo, 2012; 2016; Wehner and Thies, 2014). Emanating from national interests, domestic values, and cultural heritage, each state develops its own unique *role conception* (Breuning, 2011:28). One part of this role conception has close connection to a state's 'ego,' that is, its own belief – conception – regarding who he or she is. This ego is rather coherent no matter social context and has been described to represent the "hard core" of a state, idea-wise (see Müller, 2011). However, a state's role conception does not only consist of an ego that reviews all perceptions against an isolated 'inside.' Instead 'alter' brings in considerations taken to others as well, making role conceptions tangible to the social context. Because, when alter is added, i.e., when state A's *thoughts* incorporate also what others is believed to expect regarding its own state A's behavior, then role conceptions turn into social constructions where states start to review themselves in relation to others. International interaction and how others make use of communication in words or deeds to signal expected behavior is therefore also a fundamental element of informing states about 'who they are' (Harnisch, 2011a:8; Bengtsson and Elgström, 2011).

The second dimension of a role is constituted by *role prescriptions*. One could think of these prescriptions like the structure that surrounds interaction, informing actors of what type of behavior would be expected: ideas of which norms seem to be valid, and which behavior ought to be considered appropriate appear here. Once again others' expectations have a prominent position, but in contrast to role conceptions – where 'others' were imagined – others are *in practice* affecting role prescriptions. Actually, others' expectations are representing socializing forcers, where conformity would be the ultimate end. In order to achieve this, the social setting rises in importance, by offering an arena from where actors' senses of belonging and we-ness could be appealed to, as well as their wish for self-esteem (Wendt, 1999). We could thus think of a role as "a combination of self-conceptions and social recognition prescribed by Others" (Wehner and Thies, 2014:414). Recognition involves a normative evaluation where only some behavioral features (over others) will be recognized, and confirmed as good or bad (Iser, 2015:28). It thus represents very explicitly how prescriptions work to affect roles: an actor can act as a cooperative leader or as a cooperative laggard as much as he or she wishes, but as long as others do not recognize the state for this behavior such a position will – depending on the state preference – neither be obtained nor escaped (Bengtsson and Elgström, 2011:114). A state may therefore be forced to reevaluate its self-conception vis-à-vis others depending on how it is recognized, that is, what type of prescribed state behavior that others' recognition reveals.

The last role dimension is *role performance*, and represents the stage where the role is enacted and thereby result in an actual state behavior that is visible. The displayed performance is what is left when an actor has consulted, on the one hand, his or her ego regarding its wants and what is possibly to achieve, and on the other hand, its alter, regarding what a prescribed and appropriate behavior would be like. It is thus the sum of how the agent-structure would be like at a given situation. Even though roles show tenure in regularity (Aggestam, 2006; Backman, 1970; Holsti, 1970; Walker, 2013), the way states come to perform their roles yet leaves the (theoretical) door ajar for change. Role performances shape structure and structure shapes role performance and the imbedded between role change – expressed as learning – ultimately may come about.

### Learning as Role Change

Change in international relations originates from conflicts, not violent conflicts carried out by armed states, but smaller conflicts – those types of conflicts battled in mind. While conflicts normally would be associated with something bad like suffering, conflicts carried out in and by the mind could be something empirically good; they could induce learning, for instance, of environmental norms (see Harnisch, 2012). However, most of the times, changes because of role conflicts will rather create new conflicts and new demands for change, in an ongoing process in search of stability (Harnisch, Frank and Maull, 2011). This is due to roles being stable in representing patterned behavior and being complex having multiple expectations to cope with. Learning as role change in general signifies the process where states are made aware of how the role taken on should be performed, should a specific social position be obtained.

Stable role-playing is complicated by the fact that each state not only copes with multiple expectations, but also multiple roles. Under one big and general 'role heading,' a variety of roles are deployed – all differing depending on the issues, interactive context, and perceptions of power and security – which combined make out a full role-set. Therefore, roles cannot escape the risk of a role conflict to erupt, since role-sets include incompatibility (Aggestam, 2006:21–22). The role-set itself is a breeding ground for conflicts and these conflicts could be expected to erupt the more role incoherence there are between ego and alter expectations. Some of these conflicts could trigger role strain in self, caused by an actor being exposed to multiple expectations of various significant others active in various international institutions, thereby creating an ambiguity or contradiction between other and self-expectations (Below, 2015; Wehner, 2016). At other times, role-playing disclose incoherent expectations through role articulations compared to role performance, for instance, when actors rhetorically cherish a certain value but in action do the opposite. More concretely, to condemn coal burning in words, but in action engage in mining activities, shows a discrepancy between ego and alter expectations that could also negatively impinge on international relations by having other actors to question the legitimacy of the first actor (Bengtsson and Elgström, 2011:129). Thus, in the Arctic context, to engage in oil prospecting and extracting offshore activities, while simultaneously dedicating itself to cooperation on environmental protection, is a behavior that seems to contain contradictions. Withdrawing the Paris agreement or the previous Kyoto agreement as did the U.S. and Canada respectively, though in other respects proclaiming climate leadership, are other examples of roles that sometimes appear as irreconcilable. In situations like these, a learning process on the role could be induced, where the general 'head role' needs to be innovative to in a good way figure out which actions to favor (Aggestam, 2006:25), and how to have this packaged and presented within the role-set.

Other situations where a role conflict may cause states to review previous positions and conceptions would be vague expectations, weak knowledge of governing societal norms, and a leadership that deviates from the national role conception (Harnisch, 2012:50). Regarding the last, domestic role contestation has been revealed as an area influencing national role conceptions (Cantir and Kaarbo, 2016). As such, whereas role conflicts assume the state to play one role and have one (1) ego-attached foreign policy, "role contestation disaggregates the ego into a number of actors that contest how the state should behave" (ibid.:6). This perspective sheds light on how a national role conception, and foreign policy behavior, is not disconnected from vertical role contestations going on domestically – between elites and the public, and horizontal role contestations – between elites versus elite, i.e., between governing elites and political opposition (Cantir and Kaarbo, 2016). From this perspective, roles may not be that stable after all, but inconsistencies could spur change – for instance, through the public opinion which foresees the enactment of a different role. As the climate issue continues to climb the agenda of public opinion in country after country, international role contestations could turn out to be a key component in having politics revised, by giving new meaning to roles.

Earlier in this chapter the idea was presented that states – to a certain extent – were getting told of their wants. The perhaps most straightforward illustration of this would be alter-casting. Although not the result of a role conflict per se, it sheds light on how social recognition sometimes works to prescribe (change) behavior in international relations. Since role-playing builds on the notion of counter-roles (Walker, 2013; Wendt, 1999) alter-casting refers to the process where actors engage in negotiation to persuade others of the legitimacy of their own role. In so doing, they attempt to 'cast' others a confirmative role thereof (McCourt, 2012:380; Harnisch, 2012), with the underlying logic of having this to fit the role (vision and goal) of the first actor (McCourt, 2012). That is, a protector needs someone to protect from, a teacher needs pupils, and a leader needs someone that follows. It could be agent-driven and executed by a politically aware actor, but it could also have its source of origin in socialization, where a group of states are acting as teachers on certain norms (Harnisch, 2011a:13). In addition, Stephan Klose (2019) has argued alter-casting to not only be pivotal when actors attempt to realize their regional role aspirations, but also to affect the social structure of a region. Thus, depending on the prevalence and the need of counter roles, or the demand for fostering on norms, the Arctic region can develop through a mixture of roles taken on both by individual states and roles being casted by others.

However, from a conceptual learning perspective, states rather become aware of their wants through dedicated reflections in a given situation. A contextual change – of some kind – is expected to cause a role change, since role conflicts are more likely to erupt when conditions no longer are the same as when firstly being formulated (Aggestam, 2004:23). A new context may require changes in role performances and behavioral strategies, but an altered context my also bring about new beliefs and understandings. Role change therefore is foremost a process centered on ego and alter learning.

## Learning as a Social (Constructivist) Undertaking

Political learning has been, as described by Thies (2019), a slipper concept, with views revealing no real cohesion regarding who learns, how to study learning, and what learning is supposed to encompass. In this book, it has already been established that it is states that learn, and that role theory offers analytical entrance into such a studying thereof. What is left is thus the question of the learning content. Up until mid-1990s, there seems to be a consensus of learning as stemming from experience (Etheredge, 1981; Levy, 1994; Nye, 1987), indicating learning to be originating from lived, and past, events. With the entry of constructivism into the IR field, learning became approached in a less fixed manner, allowing for a widening of learning to concern beliefs, relational mechanisms, and norm spreading, for instance, things such as anarchy, self-help, and/or economic rationality would no longer put theoretical obstacles in the way of potential learning outcomes. Instead, learning involves cognitive processes that operate through social interaction to – from a role theoretical perspective – create visibility on how an actor's role should be enacted and performed.

An ongoing theoretical debate within IR has been whether actors base their behavior on a consequentialist logic – considering the effects of doing A or B before choosing action – or on the idea of appropriateness – i.e., if they seek to identify a "match of behavior in a given situation" (March and Olsen, 1995:30). Here, appropriateness is understood to comprise a social sensitivity that renders actors receptive to the context – its cues and demands – and ultimately to learning. Alexander Wendt has argued states to be rule-followers who are adapting to the context; they play by the rule of 'self-help anarchy' when interaction requires, but they could just as well play by the rules of friendship and solidarity (Wendt, 1992). Such a conclusion presupposes reflexivity within international relations that could involve the learning of something new, controlled for by notions of appropriateness. Finnemore and Sikkink put it in this holistic way: "[b]ecause norms involve standards of 'appropriate' or 'proper' behavior, both the intersubjective and the evaluative dimensions are inescapable when discussing norms. We only know what is inappropriate by referencing the judgement of a community or a society" (1998:891–892). As should be clear by now, roles refer to expectations or understandings of appropriate behavior (Elgström and Smith, 2006:5), why the key to unlock understandings of learning processes lies hidden in social interaction.

The quote by Finnemore and Sikkink furthermore implies that the more socialized actors are to specific norms, the more difficult it will be to have these violated. Following "the life cycle of norms" (1998), where norms firstly emerge, are cascaded into international relations, and finally are internalized within actors, socialization is a process that makes actors review their interests, positions, and capabilities against a social context that requires some form of conformity in behavior. Since such conformity is a result of norm transfers between entities (Flockhart, 2006), any learning on norms therefore presupposes that actors are interacting and therethrough influenced by fellow others (Hoffmann, 2008:148).

### *Three Kinds of Learning: Adaptive, Normative, and Ego and Alter*

In this book, three different kinds of learning are addressed; adaptive, normative, and ego and alter learning. Whereas adaptive and normative learning are result-oriented, i.e., center on *what* actors learn (strategies and values), ego and alter learning direct light to *how* actors learn. *Adaptive learning* is visible as behavioral changes, but where no reassessment of interests, behavior, or perceptions would be required. In this type of learning, new information leads to a change in means but not in ends (Levy, 1994). Therefore, it is a learning about making strategic adaptions to a context that have come to display somewhat changed rules. *Normative learning* entails a learning that targets the belief system. Others would call it complex learning (cf. Levy, 1994; Nye, 1987) or even real learning (Haas, 1990), but here it is approached as normative in order to signal its evaluative dimension; through social interaction actors have come to understand which social rules govern. As such, normative learning involves aspects of changed perceptions connected to values. Since there always is a "complex two-way interaction" (Harnisch, Frank and Maull, 2011:253–254) between actors in international relations, for instance, between states and organizational structures, the way normative learning evolves is dependent upon specific group dynamics.

Both adaptive and normative learning comprise logic of appropriateness in that this norm, in its simplest form, draws on the idea that actors try to do 'the right thing,' given context-specific actors and structures (see Risse, 2000). Also strategic decisions require knowledge of which rules – explicit and implicit – govern, and Harald Müller explains it in the following words: "[i]n negotiations, it is *appropriate* [emphasis added] for actors to pursue their self-interest unless it collides with a valid norm that prescribes a different behavior" (2004:416). Therefore, normative learning, when viewed in its simplest form acting as an informer to strategic decisions, is by no means predestined "good" or "higher" values; it simply prescribes what to count as good manners within a specific context. However, as social interaction gets deeper into actors' belief system, a normative learning may arise that encompasses a value-based improvement in relation to environmental norms. Indeed, normative learning would be a prerequisite for international learning, where the former figuratively has the shape and function of a funnel: the deeper into the funnel one gets, the more stable are beliefs, and the less room there would be for a sudden change in behavior.

*Ego and alter learning* is the mechanism, the cog, which leads to learning in various forms. Consequently, it is also the learning type of primary importance, functioning as a tool for the operationalization of learning processes. Ego and alter learning follow on role theory, arguing that learning happens through interaction, and more specifically, when actors carry out reflections on their social position in a social group. This takes place in the presence of expectations, which target interaction in four different aspects. Firstly, expectations target the ego part of an actor: who am I as an actor and what do I therefore expect of myself? Secondly, they target the alter part of an actor: how do I as an actor interpret others' expectations of me? Thirdly, expectations pertain to an actor's held expectations of others, and fourthly, expectations target the constitution of 'appropriate' behavior, for instance, in a negotiation. In more hands-on and practical terms, ego and alter learning contains an actor's interpretation of the social context given both perceived national interests and ideas, (perceived) expectations from others, and what type of behavior the cooperative mandate – explicitly through terms of reference and implicitly through norms – is prescribing. When ego and alter reflections are studied in relation to roles, learning also becomes identifiable through the observation of some kind of *role change,* caused by, for instance, a conflict or an ambiguity within actor's role conceptions (see Harnisch, 2012:53).

## Why Others Matter – Bringing the Micro Level In

To get access to the process where actors reflect on their roles, this book advocates the use of a micro perspective. More specifically, the micro perspective offers an entry into the domain of state learning by opening for two important discussions concerning consensus-reaching processes. The first discussion is conceptual and concerns how state representatives come to reflect – as well as pay attention to – expectations on the type of behavior, including those expectations stemming from others. The second one is more empirically oriented and deals with how state representatives come to view specific (scientific) claims – environmental claims – as being valid or not. These discussions draw on the work of social psychologist George Herbert Mead and sociologist Jürgen Habermas.

### *Mead and the Power of Expectations*

A core assumption within role theory is the capability of alter expectations of constraining, or forming, state behavior through what would be a mix-ture of soft and social power (Bengtsson and Elgström, 2011; Harnisch, 2012; Harnisch, Bersick and Gottwald, 2015). The question then arises: why should actors even bother to care for *others'* expectations? An early twentieth century social psychologist, the American George Herbert Mead, probes a plausible explanation through a key statement of his: "we must be others if we are to be ourselves" (Mead, 1925:276). Conformed to mutuality, the phrase entails social interaction to revel our inner selves, *if* attention is being paid to alter expectations. In other words, only by thinking through others, actors could be fully aware of their roles and wants.

The above statement is a cornerstone of the symbolic interactionism theory, where actors are understood to interpret the world through inter-action, while having these interpretations shape their behavior. Still, actors are not weak-willed, i.e., lacking a will or behavioral direction of their own. Instead state behavior – and likewise roles – is a combined result of social impact and individual convictions. The appeal to role theory of Mead's way of reasoning has been observed before. His approach has particularly been considered helpful in relation to 'role-taking' processes within inter-national relations[1] (Beneš and Harnisch, 2015; Harnisch, 2011b; 2015; Klose, 2019; McCourt, 2012). Mead illustrates the way that actors become aware of their role with the help of a play-and-game analogy, which really is an analogy on socialization. Play and game represent different levels of self-consciousness: rationality *and* impersonal reasons are, for instance, equally important for an actor's ability to obtain self-consciousness (Mead, 1934:138). In 'play,' where the analogy starts off, children are continuously imitating the roles of others: the small child shows awareness of society and starts to role-play as mom and dad, as crook and hero, as doctor and cashier staff, and other roles found in the society to which the child belongs. However, the child is not bigger than his or her lived experiences, and role-playing responses are therefore limited in scope. When the child grows older, he or she instead turns to playing games. A game involves following rules and it involves a regulated behavior, thus demanding a greater imaginary ability regarding society: by taking on the roles of what Mead refers to as "generalized others" – an abstract social category pertaining to, for instance, what is means to be 'human' or part of a certain group – the child no longer takes on the role of only specific (or significant) others (like the crook, hero, or doctor), but of *all others* involved (Beneš and Harnisch, 2015). By anticipating the attitudes and expectations of many, the world grows to the child but more importantly he or she now will be provided with a Self; by responding and reacting through others, he or she becomes a social object (Mead, 1925:269; 1934:153–154).

But what then has all of this to offer a study of learning? Well, if one, like Mead, dissects an actor's self into *I* and *Me,* then there is more analytical clearance brought into the response offered by social stimuli. Interacting actors are constantly engaged in internal dialogues between *I*s (a rather irreducible part of self where the ego part of a role hosts) and *Me*s (where others' expectations are interpreted). Together *I* and *Me* constitute an actor's *Self.* Given its context-specified character, the Self is like an in situ role. When an actor that is a beginner on chess, football, or a card game sticks to it and plays one game after another, the actor will eventually be socialized in all those rules governing the game, how to behave strategically in relation to others, and what a sportsmanlike behavior would be. In a stable situation, for instance, when the same game is played repeatedly, *Me* works in a routine manner and sticks to the norms established by social practice (Mead, 1934; Harnisch, 2011a:10–11). Should then a 'problematic situation' of some kind occur (Harnisch, 2011b; Klose, 2019), the rules would no longer have the same significance why actors potentially are fostered to learn something new. Such a learning process starts with *I* – which is more impulsive and thus keener to change – having taken precedence over *Me* and stepped outside of the routine situation (Mead, 1934:210–213). It would, however, not count as learning since it would happen rather unreflective and unaware: *I* is mainly searching for freedom from a social situation primarily controlled by *Me* (ibid.:176–177). The potential for learning would thus enter only the picture through *Me,* when it takes in the community and their attitudes, and starts to a norm change with *I*. Others' expectations, again, thus matter.

The micro perspective offered by Mead presents a hypothesis of how expectations are processed within the belief system of actors. Since it furthermore acknowledges actors' reflective skills, Mead sees ahead a potential progress in human interaction. The formula is to pay greater attention to others' interests:

> I think all of us feel that one must be ready to recognize the interests of others even when they run counter to our own, [and] that the person who does that does not really sacrifice himself, but becomes a larger self.
>
> (Mead, 1934:386)

To GEG, the recipe for a social outcome such as international learning on environmental norms thus includes socialization on those values that support the self as a social being, rather than an individualist ditto. As will be shown next, communication here fills an important function.

### Habermas and the Importance of Communication

Owing the potential for change that is being inherited in communicative action, others before have discussed the importance of Jürgen Habermas to IR theory in general and role theory in particular (see, for instance, Deitelhoff and Müller, 2005; Diez and Steans, 2005; Müller, 2011). Habermas, a German sociologist, reveals actors to be reflective, longing for informed decisions, and his theory of communicative action offers input on how expectations – and above all how *alter* expectations that originally operate from the outside-in – become known to actors in the first place, therethrough possible to consider.

To Habermas, communication and the transmitting of information are fundamental for any interacting actor. Relying on communicative action, Habermas describes human interaction as oriented toward common understanding (1984). Through a common understanding, people are armed with an attitude toward interaction where coordination of actions seems obvious and self-explanatory: due to an exchange of interpretations and views of the world, the communicative process has made actors synchronized in behavior (ibid.). For GEG and an international learning perspective, such a common understanding of course would be valuable. In the Arctic Council, the Icelandic Minister for Foreign Affairs offered a verbalized example of the existence of common understanding in the AC, pointing toward the Arctic as of common ground and common responsibility:

> We may not all agree on every single issue, but I do dare to say that we do share a *common understanding* [emphasis added], namely, that the Arctic is an important region not only to all of us present here, but also in the context of global environmental development.
>
> (Thórdarson, 2019)

However, although it could be that the Arctic states share an overarching common understanding of the Arctic region in practice, far from all communication leads to a common understanding on issue-specific matters like environmental norms. Indeed, should communication lead to an understanding that would halter in commonness, which often is the case in negotiations, a conflict would arise leaving actors with three choices: turn toward individualist strategic action, walk away from interaction, or turn to discourses in order to establish what is true and what is false. Only if the last is chosen, actors are still committed to communicative acts by engaging in the discursive practice of searching 'the better argument,' i.e., justified understandings for everyone to accept. What Habermas says is that consensus is something not only to be reached through compromises, but also that arguments matter: good and convincing arguments contain a power of persuasion.

To engage in a process of communicative action serves to test specific (scientific) claims in terms of their validity, i.e., to establish the discourse on specific issues. Within the Arctic Council and amongst the Arctic states, the discourse surrounding climate change – if by climate change we draw connections between greenhouse gases predominantly such as carbon dioxide and a warmer climate – has, for instance, remained an issue to be addressed on a global scale through UN regimes. In an Arctic context, *actions* on climate change have instead centered on a reduction of short-lived climate pollutants, such as black carbon and methane. The Arctic climate change discourse seems thus fit a political context where oil and gas ensure economic and social development, prior that of being environmental threat. However, to return to the process of testing validity claims, these claims, and thus the understanding sought for, relate to the world in three different ways: (1) the statement made by an actor has to be true (in the objective world); (2) it has to be considered right in a normative context (in the social world); and (3) the intention of the speaker and the speech act must be honest (in the subjective world) (Habermas, 1984:99). To simplify, do actors share the same beliefs regarding arrangements of the world, do they share norms, and do they trust each other to be sincere in what they are saying? The better, more transparent, and long-run the arguments, the more likely it is that all of the members will be convinced of its validity, with good action on environmental protection subsequently achieved (ibid.: 42).

What is appealing with Habermas' way of reasoning is the inherited promise of change; if actors are searching for good arguments, they are likewise open for reevaluations of the mind, i.e., to learn. However, following Müller (2011), the likelihood for communicative action to take place stands in stark connection with the type of agency (character of actor, their roles) and the degree to which their interactive setting is institutionalized. According to him, states that follow a script of moral entrepreneurs, and who operate within a regulated and dense institutional environment, would – compared to rogues and an unmitigated anarchical structure – be more inclined to engage in communicative action. For instance, within the Arctic Council a state such as Russia tends to be described to cause communicative dubiousness through a negotiation style being rather difficult to understand.[2] "To negotiate with Russia is *always* difficult" a delegate with extensive AC experience concluded (Interview, 2014). Communication with Russia – and understandings thereof – are likely colored by Russia's political structure, where it deploys an authoritarian political system while also striving to reemerge as a superpower. At the same time, interview accounts concerning Russia give rise to positive judgments from fellow negotiators, where the state has turned out to be – as perceived by others – more inclined toward cooperation than expected. Thus, Arctic Council cooperation reveals eight state parties ready to communicate, possibly ready to change or even learn.

### *Learning as Arguing, Arguing as Learning*

For states performing roles, language and communication effect learning by laying bare expectations, norms, and prescriptions for the role to react on. In the Arctic Council, the development of the *Fairbanks Declaration* from 2017 is one example of the political power that language possesses, used to escape unsolicited role expectations in future cooperation. Because, when the U.S. chaired the AC Ministerial meeting and confirmed a continuous commitment to protect the fragile Arctic environment including climate change, it was yet an issue that was getting down-prioritized by the political administration. A leaked draft of the meeting declaration also showed how the U.S. just the day before had required six changes to the declaration, where each and every one of them methodically weakened the language on climate change and climate change action (Shankman, 2017). By weakening the language, cooperative expectations were sought to be lowered, and in parallel the ambition for common understanding on (Arctic) climate change would go down.

Language is a powerful tool that could both attempt to resist change or promote change. As Habermas (2005:249) explains, facts do seldom speak for themselves:

> Facts can be explained only by resource to factual statements, what is real only by appeal to what is true. Since the truth of beliefs and sentences in turn can be justified or disputed only by means of the beliefs and sentences, we – as reflecting agents – cannot step outside the circle of language.

From this perspective, international negotiations on environmental claims – like those found in the Arctic Council – can be approached as intersubjective truth-seeking processes, operating via language, and with analytical focus directed at interpersonal relations.

However, despite international negotiations at large being sprung from dynamic social processes, the negotiation literature has been criticized for neglecting such relational dimension (see, for instance, Gelfand et al., 2006; Koc-Menard, 2009; Jönsson, 2015). Scholars have therefore argued for the relevance of including also social norms and relational aspects in the study of negotiations, where, for instance, Schoppa (1999) explicitly has referred to norms as what defines notions of "legitimacy" within legitimate bargaining tactics, being partly influenced by level of trust among parties. Relationships between negotiators, operating in context-specific social environments, seem furthermore to be of increasing importance as international negotiations are getting more permanent, more focused on solving common global problems like environmental pollution, and often preceded by specialized agencies like epistemic communities who share beliefs and values across state borders (Jönsson, 2015). Within such a setting, just as interpersonal relations should not be neglected, neither should the power of language – intersubjective structures of meaning and the interpretations of words and symbols – be underestimated (ibid.:18).

From this perspective, *arguing* becomes an analytical concept that is accessed through interpersonal relations at the micro level, and which unfolds the potential for change in international relations. Thomas Risse has explained arguing to govern actor's decision-making process and to be a distinct model for social interaction, which includes process of argumentation, deliberation, and persuasion. "Apart from utility-maximizing action, on the one hand, and rule-guided behaviour, on the other, human actors engage in truth seeking with the aim to reach mutual understanding based on reasoned consensus, challenging the validity claims involved in any communication" (Risse, 2000:1–2). Arguing is furthermore an implicit learning mechanism, against which actors not only do evaluate interests, knowledge, and information, but where they also reflect and assess the validity of norms (2004:288). Moreover, it is an illustration of Habermas' concept of discourse. Adapting Habermas' sociological theory to an IR

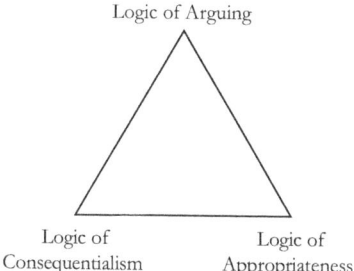

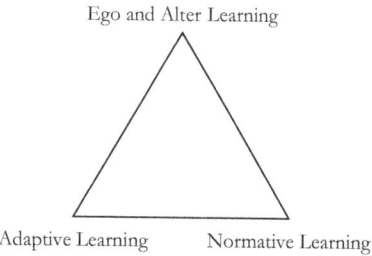

*Figure 2.1* Three Logics of Social Action.
Source: Thomas Risse (2000:4).

*Figure 2.2* Three Kinds of Learning in Social Interaction.
Source: Modified by author.

context, Risse has developed the triangle *Three Logics of Social Action* (2000). This model provides a scheme for how communication reveals to actor which type of action-based logic to apply, for instance, when engaged in negotiations (Figure 2.1). The contribution lays in presenting arguing as a via media logic – a bridge – between logic of consequentialism and logic of appropriateness (2000:2004).

A modified version of Risse's three logics of social action is presented in Figure 2.2, *Three Kinds of Learning in Social Interaction,* now with learning processes made explicit; here, interaction is triangularly depicted with the process located on top and the outcome at the base. The upper corner of the triangle acts as the driver of movement, as in Risse's, the cog that keeps the process going. Just like arguing decides on which logic to apply, ego and alter learning constitutes the mechanism that pushes actors in the direction of either adaptive or normative learning, and is where the book has its focus. Whereas adaptive learning is more of a contextual reaction, rapid in its response, deep learning on norms and values is more profound.

A micro perspective on interpersonal relations, like those found in negotiations, gives input on the consensus-reaching process where actors attempt to render coordination of actions possible, relying on arguing. To reach a good result, there should be a common understanding on what the language is signifying, and the conveyed values of the very same language. For instance, as in the case of Arctic oil pollution prevention, to *prevent* oil pollution could mean to actually make sure a spill will never ever happen, using all means at one's disposal. Such an understanding would, however, first have to go through a communicative process establishing the discourse on oil pollution to prevention to precisely mean this, and not just an improvement on oil safety procedures aiming for risk reduction. To argue in such a way that common understanding on the former definition of prevention would reign requires a process where some actors are being normatively persuaded to learn to think differently regarding 'facts of reality.'

## A Model for the Correlation between Learning and Roles

The idea put forward is to view the learning process in three interrelated steps: firstly, as Martha Finnemore stated (1998), states do not always know what they want. Yet, states still end up wanting something. Socialization, operating through a structure containing appropriateness, expectations, and recognition, here has a 'telling-effect,' by teaching states of their wants. Secondly, states do something, i.e., they take on a role and engage in role-playing. However, how the role should be played is something to be determined in relation to the social structure, and what the actor considers to be an appropriate behavior given both ego and alter expectations. Thirdly and lastly, the role performance of states is a product of alter and ego reflections, which is analytically accessed through a micro perspective and which furthermore has the potential to activate state learning.

It is now time to summarize this into a model, or hypothesis, for how learning gets manifested within roles, and thus, how learning processes fruitfully may be studied (Figure 2.3). This model illustrates arguing to be the engine for learning, located as it is in the center of the triangle. By arguing, actors attempt to settle contestations on norms, to arrive at a stage of reasoned consensus. Arguing here embodies expectations, which are fed into the process from the outside through things such as traditional knowledge, scientific findings, rules, practices, and so on, thereby keeping the process in movement. The arrows that circulate the triangle represent understanding to be an ongoing process, where interpretations of the social context (roles and norms) are under constant change. The clockwise direction signals that also strategic and unilateral behavior has its start in considerations of whether such behavior would be contextually 'appropriate' or not.

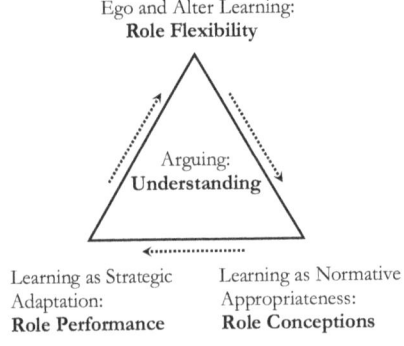

Figure 2.3 The Correlation between Learning and Roles.

The model links learning to different components – or expressions – of the role. At the top of the triangle, ego and alter deliberate and reflect on all things they come to know of, in terms of expectations, prescriptions, interests, and self-conceptions. Through a 'rough calculation' it will then identify the most beneficial social position to obtain. This process of ego and alter learning, described as a "complex two-way interaction" between states' roles and organizational structures (Harnisch, Frank and Maull, 2011:253–254), takes the expression of a role flexibility. There is a certain amount of flexibility within roles that allow them to socially respond to the context, emphasizing different dimensions of its role-set, without losing grip of who they are.

The model continues by having adaptive learning and normative learning each coupled with one separate role component: it is suggested that adaptive learning couples well with a change in role performance, whereas (profound) normative learning *also* gets reflected in role conceptions. Indeed, the observant reader would here want to raise an objection; would not role performance be a combined sum of multiple behavioral estimates that have been carried out, and why it therefore would simply not be possible to disconnect normative learning from the performative part of the state role? This reader would be right in his or her observation: an actor that has learned something normatively new would most likely have this reflected in role performance. However, what the triangle attempts to depict, with its distinction between learning and role performance, on the one hand, and learning and role conception, on the other hand, is that adaptive and normative learning are different when it comes to depth of understanding. An actor can for strategic reasons choose to behave in accordance with environmental norms without necessarily *understanding –* sharing the belief of – these values as true: to learn in an adaptive way does not require any profound changes in role conceptions, especially not ego one, but suffice would be to be somewhat attentive to prescriptions. Normative learning is for sure also part of adaptive learning, by making actors aware of new prescriptions to obey by (although not necessarily believe in). It is, however, also a kind of learning that as it moves down the 'funnel of understanding' eventually may arrive at a thick learning. At such an ultimate stage, new norms will eventually be fully absorbed by the actor's ego and alter conceptions, and thus forming part of an actor's *I*.

### Summary

In this chapter, a theoretical framework has been presented, designed for an analysis of learning through roles. It has been discussed how a separation of an actor's role into an ego part and an alter part, and between role conceptions, role prescriptions, and role performances, reveals how roles are potentially carriers of incoherence. Such incoherence contains conflicts that could lead to change and learning. With help from symbolic interactionism and G. H. Mead, a micro perspective adds insight on how actors deliberate on these potentially incoherent parts. In addition, communication – or rather arguing – represents the persuasive tool that actors use to reach a consensus on how roles should be played, and empirical problems solved, or at least, approached.

The chapter has suggested the possible effect of role-playing to be a learning that gains visibility in role performances, or possibly in role conceptions should it go deep into beliefs and cause alterations to (ego) conceptions. Ultimately, the model for the correlation between learning and roles suggests *that any learning on environmental norms, firstly runs through a social learning on roles,* where the role is attempting to figure out how best to be presented while yet 'made to fit' a context encapsulating new environmental demands. For a similar reason is learning on environmental norms – that stretches beyond strategic reasoning on to the belief system – also likely to take time.

## Notes

1　Roles in international relations evolve through three stages: role-taking, role-making, and alter-casting. Role-taking refers to the process where roles are played out – enacted – in international relations, but where the enactment itself causes states dubiousness regarding what is embedded in the role in terms of expectations. Role-making would instead be little more of an agent-focused process; here, states are attempting to initiate action (on the role taken on) and have it accepted by Others. Alter-casting would, as described elsewhere in this thesis, be the act of casting Others into roles with the underlying logic of having this to fit the role (vision and goal) of the first actor (McCourt, 2012:379–380).

2　The view of Russia as being communicative difficult to understand was a recurrent theme in many of the interviews carried out in this book. The communicative difficulty concerned both practical and unintentional misunderstanding stemming from language use, and the negotiation style per se where few words were uttered. A further complicating factor was national institutions and structures, understood to constrain Russian negotiators' room to maneuver.

## References

Aggestam, L. (2006). "Role Theory and European Foreign Policy: A Framework of Analysis," in Elgström, O. and Smith, M. (eds.). *The European Union's Roles in International Politics: Concepts and Analysis.* London and New York: Routledge.

———— (2004). *A European Foreign Policy? Role Conceptions and the Politics of Identity in Britain, France and Germany.* Stockholm Studies in Politics 106. Stockholm: Stockholm University.

Akbaba, Y. and Özdamar, Ö. (2019). *Role Theory in the Middle East and North Africa: Politics, Economics and Identity.* New York: Routledge.

Backman, C. (1970). "Role Theory and International Relations: A Commentary and Extension," *International Studies Quarterly,* Vol. 14:3, pp. 310–319.

Below, A. (2015). *Environmental Politics and Foreign Policy Decision Making in Latin America: Ratifying the Kyoto Protocol.* London and New York: Routledge.

Beneš, V. and Harnisch, S. (2015). "Role Theory in Symbolic Interactionism: Czech Republic, Germany and the EU," *Cooperation and Conflict,* Vol. 50:1, pp. 146–165.

Bengtsson, R. and Elgström, O. (2011). "Reconsidering the European Union's Roles in International Relations: Self-Conceptions, Expectations, and Performance," in Harnisch, S., Frank, C. and Maull, H. W. (eds.). *Role Theory in International Relations: Approaches and Analyses.* London and New York: Routledge.

Breuning, M. (2011). "Role Theory Research in International Relations: State of the Art and Blind Spots," in Harnisch, S., Frank, C. and Maull, H. W. (eds.). *Role Theory in International Relations: Approaches and Analyses.* London and New York: Routledge.

Cantir, C. and Kaarbo, J. (2016). "Unpacking Ego in Role Theory: Vertical and Horizontal Role Contestation and Foreign Policy," in Cantir, C. and Kaarbo, J. (eds.). *Domestic Role Contestation, Foreign Policy, and International Relations.* London and New York: Routledge.

———— (2012). "Contested Roles and Domestic Politics: Reflections on Role Theory in Foreign Policy Analysis and IR Theory," *Foreign Policy Analysis,* Vol. 8:1, pp. 5–25.

Checkel, J. T. (2003). " 'Going Native' in Europe? Theorizing Social Interaction in European Institutions," *Comparative Political Studies,* Vol. 36:1/2, pp. 209–231.

———— (2001). "Why Comply? Social Learning and European Identity Change," *International Organization,* Vol. 55:3, pp. 553–588.

Deitelhoff, N. and Müller, H. (2005). "Theoretical Paradise – Empirically Lost? Arguing with Habermas," *Review of International Studies,* Vol. 31:1, pp. 167–179.

de Sá Guimarães, F. (2020). *A Theory of Master Role Transition: Small Powers Shaping Regional Hegemons.* New York: Routledge.

Diez, T. and Steans, J. (2005). "A Useful Dialogue? Habermas and International Relations," *Review of International Studies,* Vol. 31:1, pp. 127–140.

Elgström, O. and Smith, M. (2006). "Introduction," in Elgström, O. and Smith, M. (eds.). *The European Union's Roles in International Politics. Concepts and analysis.* London and New York: Routledge.

Etheredge, L. D. (1981). "Government Learning: An Overview," in Long, S. L. (ed.). *The Handbook of Political Behaviour,* Vol. 2. New York: Plenum Press.

Finnemore, M. (1996). *National Interests in International Society.* Ithaca: Cornell University Press.

Finnemore, M. and Sikkink, K. (1998). "International Norm Dynamic and Political Change," *International Organization,* Vol. 52:4, pp. 887–917.

Flamm, P. (2019). *South Korean Identity and Global Foreign Policy: Dream of Autonomy.* New York: Routledge.

Flockhart, T. (2006). "Complex Socialization: A Framework for the Study of State Socialization," *European Journal of International Relations,* Vol. 12:1, pp. 89–118.

Gelfand, M. J., Smith Major, V., Raver, L. J., Nishii, L. H. and O'Brien, K. (2006). "Negotiating Relationally: The Dynamics of the Relational Self in Negotiations," *Academy of Management Review,* Vol. 31:2, pp. 427–451.

Haas, E. (1990). *When Knowledge Is Power: Three Models of Change in International Organizations.* Oakland, CA: University of California Press.

Habermas, J. (2005). *Truth and Justifications.* Cambridge, MA: MIT Press.

——— (1984). *The Theory of Communicative Action. Reason and the Rationalizaton of Society,* Vol. 1. Cambridge: Polity Press.

Hajer, M. A. (2005). "Setting the Stage. A Dramaturgy of Policy Deliberation," *Administration & Society,* Vol 36:6, pp. 624–647.

Harnisch, S. (2012). "Conceptualizing in the Minefield: Role Theory and Foreign Policy Learning," *Foreign Policy Analysis,* Vol. 8:1, pp. 47–69.

——— (2011a). "Role Theory: Operationalization of Key Concepts," in Harnisch, S., Frank, C. and Maull, H. W. (eds.). *Role Theory in International Relations: Approaches and Analyses.* London and New York: Routledge.

——— (2011b). "'Dialogue and Emergence': George Herbert Mead's Contribution to Role Theory and His Reconstruction of International Politics," in Harnisch, S., Frank, C. and Maull, H. W. (eds.). *Role Theory in International Relations: Approaches and Analyses.* London and New York: Routledge.

Harnisch, S., Bersick, S. and Gottwald, J-C. (2015). "China's International Roles," in Harnisch, S., Bersick, S. and Gottwald, J-C. (eds.). *China's International Roles: Challenging or Supporting International Order?* Florence: Taylor and Francis.

Harnisch, S., Frank, C. and Maull, H. W. (2011). "Conclusion: Role Theory, Role Change, and the International Social Order," in Harnisch, S., Frank, C. and Maull, H. W. (eds.). *Role Theory in International Relations: Approaches and Analyses.* London and New York: Routledge.

Hofmann, C. (2008). *Learning in Modern International Society. On the Cognitive Problem Solving Abilities of Political Actors.* Wiesbaden: VS Verlag für Socialwissenschaften.

Holsti, K. J. (1970). "National Role Conceptions in the Study of Foreign Policy," *International Studies Quarterly,* Vol. 14:3, pp. 233–309.

Iser, M. (2015). "Recognition between States? Moving beyond Identity Politics," in Daase, C., Fehl, C., Geis A. and Kolliarakis, G. (eds.). *Recognition in International Relations: Rethinking a Political Concept in a Global Context.* Basingstoke: Palgrave Macmillan.

Jönsson, C. (2015). "Relationships between Negotiators: A Neglected Topic in the Study of Negotiation," *International Negotiation,* Vol. 20:1, pp. 7–24.

Keohane, R. O. (1984). *After Hegemony: Cooperation and Discord in the World Political Economy.* Princeton and Oxford: Princeton University Press.

Klose, S. (2019). "The Emergence and Evolution of an External Actor's Regional Role: An Interactionist Role Theory Perspective," *Cooperation and Conflict,* Vol. 54:3, pp. 426–441.

Knopf, J. W. (2003). "The Importance of International Learning," *Review of International Studies,* Vol. 29:2, pp. 185–207.

Koc-Menard, S. (2009). "How Negotiators Can Use Social Relations to Create Advantage," *Human Resource Development International,* Vol. 12:3, pp. 333–341.

Levy, J. (1994). "Learning and Foreign Policy: Sweeping Conceptual Minefield," *International Organization*, Vol. 48:2, pp. 279–312.

Malici, A. and Walker, S. G. (2016). *Role Theory and Role Conflict in U.S.-Iran Relations: Enemies of Our Own Making.* New York: Routledge.

March, J. G. and Olsen, J. P. (1995). *Democratic Governance.* New York: Free Press.

McCourt, D. M. (2012). "The Roles States Play: A Meadian Interactionist Approach," *Journal of International Relations and Development*, Vol. 15:3, pp. 370–392.

Mead, G. H. (1934/1972). *Mind, Self and Society. From the Standpoint of a Social Behaviourist.* Edited and with an introduction by C. W. Morris. Chicago: University of Chicago Press.

——— (1925). "The Genesis of Self and Social Control," *International Journal of Ethics*, Vol. 35:3, pp. 251–277.

Müller, H. (2011). "Habermas Meets Role Theory: Communicative Action as Role Playing?" in Harnisch, S., Frank, C. and Maull, H. W. (eds.). *Role Theory in International Relations: Approaches and Analyses.* London and New York: Routledge.

——— (2004). "Arguing, Bargaining and All That: Communicative Action, Rationalist Theory and the Logic of Appropriateness in International Relations," *European Journal of International Relations*, Vol. 10:3, pp. 395–435.

Nye. J. S. (1987). "Nuclear Learning and U.S.-Soviet Security Regimes," *International Organization*, Vol. 41:3, pp. 371–402.

Risse, T. (2004). "Global Governance and Communicative Action," *Government and Opposition*, Vol. 39:2, pp. 288–313.

——— (2000). " 'Let's Argue!': Communicative Action in World Politics," *International Organization*, Vol. 54:1, pp. 1–39.

Shankman, S. (2017). "Leaked Draft Shows How U.S. Weakened Climate Change Wording in the Arctic Declaration." *Inside Climate News*, May 19.

Schoppa, L. J. (1999). "The Social Context in Coercive International Bargaining," *International Organization*, Vol. 52:2, pp. 307–342.

Sterling-Folker, J. (2000). "Competing Paradigms or Birds of a Feather? Constructivism and Neoliberal Institutionalism Compared," *International Studies Quarterly*, Vol. 44:1, pp. 97–119.

Thies, C. G. (2019). "Political Learning and Socialization," *Oxford Bibliographies.* Last modified March 27, 2019, accessed via oxfordbibliographies.com.

——— (2014). *The United States, Israel and the Search for International Order: Socializing States.* New York and London: Routledge.

——— (2010). "Role Theory and Foreign Policy," in Denemark, R. A. (ed.). *The International Studies Encyclopedia.* Blackwell, Blackwell Reference Online.

Thies, C. G. and Breuning, M. (2012). "Integrating Foreign Policy Analysis and International Relations through Role Theory," *Foreign Policy Analysis*, Vol. 8:1, pp. 1–4.

Thórdarson, G. T. (2019). *Speech by the Minister of Foreign Affairs of Iceland.* AC Ministerial meeting, Rovaniemi. Member statement. Available via artic-council. org.

Walker, S. G. (2013). *Role Theory and the Cognitive Architecture of British Appeasement Decisions. Symbolic and Strategic Interaction in World Politics.* New York and London: Routledge.

Wehner, L. E. (2016). "Inter-Role Conflict, Role Strain and Role Play in Chile's Relationship with Brazil," *Bulletin of Latin American Research*, Vol. 35:1, pp. 64–77.

Wehner, L. E. and Thies, C. G. (2014). "Role Theory, Narratives, and Interpretation: The Domestic Contestation of Roles," *International Studies Review*, Vol. 16:3, pp. 411–436.

Wendt, A. (1999). *Social Theory of International Relations*. Cambridge: Cambridge University Press.

——— (1992). "Anarchy Is What States Make of It: The Social Construction of Power Politics," *International Organization*, Vol. 46:2, pp. 39–425.

# 3 Above the Timberline – The Logic of Arctic Cooperation

## Introduction

At a latitude of approximately 66 degrees north (66°32'N) of the equator runs the Arctic circle.[1] According to the scientific definition, once this latitude has been crossed, the polar world has been entered. From this circle, departing in Sweden, the air distance to the Norwegian North Cap – often considered the northern most point of the European mainland – measures less than 700 km. From the North Cap, an additional 2,100 kilometers must be trekked before the North Pole could be proclaimed reached.

The Arctic is a vast area, a huge geographic area that is home to four million people. The region comprises cities, even big ones, like the Russian city of Murmansk which in its heydays was home to half a million people. However, for most people the Arctic does not bring cities, cars, or shops to mind. Instead, it is thought to be a remote area – harsh, icy, and cold – triggering the adventurous and explorative nerve in those who from the outside are looking in. During the last few centuries, adventurous expeditions 'to become the first man ever'[2] have succeeded each other, in man's longing for outwitting nature's brute force. Today, also, adventurers continue to head to the Arctic, but instead of ships they often come alone, skiing, hiking, or canoeing, in what could be more modest attempts of becoming one with nature. The idea, however, of the Arctic nature as something to master still clingers on, now with economic connotations: cruise tourism – where tourists are allured to travel in the footsteps of great explorers and go on a Northwest Passage voyage or embark an icebreaker headed the North Pole – new cost- and time-saving shipping lines, and the extraction of Arctic natural resources are all contemporary illustrations of attempts to, in a way, outsmart nature in a region that previously has been known for its inaccessibility.

The fact is that the Arctic is changing. The climatic changes undergone over the past fifty years have been profound: inevitable, the region is heading a future where its environment will be warmer, wetter, and more variable (AMAP, 2017:3). Over the last few decades, at the close of every winter, it was declared that never before in our modern history has the global average temperature been as high, the Arctic as warm, its sea-ice as thin, and its spreading as limited.[3] For the majority of those approximately 500,000 indigenous people who live in the Arctic, often in a symbiotic relationship with nature, this is a worrisome development that affects everyday life. Arctic warming impedes a traditional way of living, for instance, for people dependent on ice for reaching hunting areas, or for those who find the thawing soil to lack sufficient bearing capacity for homes and settlements (ibid., 2017).

At the same time, for the Arctic states, an Arctic region that is getting warmer has developed into a strategic resource base with increasing potential for businesses and states alike. The intertwined relationship between energy interests on the one hand and environmental protection on the other is there for the Arctic states to relate to, also when cooperating. This chapter discusses Arctic development as a process informed by forces calling for change, and forces that attempts to resist. Tensions and conflicts originating from competing interests have not been insignificant, neither throughout prevailing. It describes Arctic state cooperation, as it for more than two decades has been governed under – and developed through – the Arctic Council, where it shows Arctic state interaction to proceed with a direction and cooperation to progress.

## The Arctic Drama

The Arctic Council is an institutionalized continuation of the AEPS, agreed to by the eight Arctic states in the city of Rovaniemi in 1989, and signed two years later. The legacy of AEPS is noticed within present day AC in four different ways: it recognized the Arctic region as valuable and expressed concern over threats to its environment; it encouraged scientific work by calling for assessments of environmental impact from human activities, including climate change and oil pollution; it foresaw cooperation as the only viable path forward, and WGs and indigenous representation were considered key to provide states with important input for making decisions.

AEPS is often referred to as the 'Finnish initiative,' due to it being Finland who organized and invited states to the Rovaniemi meeting. The initiative was clearly located in time and space, where the Chernobyl accident in 1986 was an abrupt awakening concerning Arctic interdependence and a triggering factor for regional environmental protection (Nord, 2016a). Indigenous people and Arctic ecosystems were badly suffering from its radioactive contamination, and today, still, reindeer pastorals must check each reindeer for levels of caesium-137 before the animal can serve as food.

There is, however, more to Arctic cooperation than caring for the environment, which predominantly rests upon a wish for peace and stability. Such an objective goes back to the end of the Cold War era and Mikhail Gorbachev's Murmansk speech. In his speech, he declared the heavy militarized Arctic region – perhaps a bit surprisingly – to represent an opportunity to have it transformed into a "zone of peace" (Gorbachev, 1987). Whereas the importance of the Arctic as a nuclear-free zone was clear from a peace perspective, it was equally important from an environmental perspective. The legacy of the nuclear arms race had heavily permeated the Arctic region from nuclear reactors, accidents, and waste disposal.[4]

In the Arctic setting, perhaps environmental protection should not be considered a freestanding issue but as part of a broader political agenda aiming for regional constructiveness (Nilsson, 2012:181). Indeed, the Murmansk speech could be viewed as a momentum for the building of the Arctic as a region, made more explicit through interaction (Keskitalo, 2004:42). However, as pointed out by Keskitalo, building a region is not a coherent task, but one where different actors, interests, and perceptions of what the Arctic 'is' yet is put under the same Arctic cooperative umbrella. By not problematizing the Arctic as a region, conflicts (of interests/between actors) become unsolvable (ibid., 177). In that sense, that the Arctic region and Arctic relations stay characterized by cooperation is by no means a self-fulfilling prophecy, but a matter of political choice.

In the Arctic, the strife for environmental protection exists side by side – and not always supportively – with the harvesting of material interests. As such, Arctic relations represent two sides of the same coin; the environmental side may be filled with incentives for cooperation but flipping the coin to the other side reveals unilateral and strategic thinking *en masse*, foremost depicted as the strife for natural resources. Had the Arctic been all about material interests, it could have been a drama of states making aggressive moves in the name of sovereignty. It *could* have, but it is not.

*Not a Drama on Major Conflicts…*

The Arctic region is commonly referred to as 'the last resource frontier,' not least does it pertain to Arctic oil and gas. States like Canada, Norway, Russia, and U.S. did start their offshore extraction activities many decades ago, where Russia and Norway even have established themselves as global top producers, much thanks to Arctic discoveries. In 2008, the U.S. Department of Interior published an assessment of remaining oil and gas resources in the world, conducted by the U.S. Geological Survey (USGS). The Arctic was declared the winner, estimated as storing 20–25 percent of all yet undiscovered hydrocarbon reserves in its sediments. Given its message, perhaps it comes as no surprise that the assessment has had a huge impact, referred to by businesses, politicians, and academia. Really, the USGS publication did more than present figures and resource estimates, it benchmarked a political shift, or rather more possible, a new Arctic rhetoric.

Coalescing in time with both USGS and reports on detracting sea ice, the Arctic was described as a region becoming increasingly centered on conflicts and rivalry (Borgerson, 2008; Posner, 2007). Territorial claims pertaining to land in the Arctic were and remain close to settled, but several overlapping claims still exist in relation to the sea, or more precisely, over the seabed storing resources like oil and gas (Byers, 2009; 2013; Dodds and Nuttall, 2016). Although not the only one, the most symbolic example of an ownership issue yet to be resolved is the claims by Denmark, Russia, and more recently Canada, to the Lomonosov ridge stretching over the North Pole. Relying on scientific investigations to support their respective case, the ridge is argued a natural prolongation of territorial continental shelves. Should any of these claims be approved by the UN Commission on Delimitations of Continental Shelves, that state would be given the exclusive right to extract any resources found in the seabed or subsoil.

With not all the borders fixed, the risk for geopolitical conflicts could *potentially* be expected (Runge Olesen, 2015). Still, Arctic relations did not, and has not, evolved into a drama on major conflicts. A vast number of scholars have come to devalue the risk of conflicts breaking out in the Arctic (Byers, 2013; Chater, 2016; Exner-Pirot, 2016; Heininen, 2015; Hough, 2013; Keil, 2014; Nicol and Heininen, 2014; Young, 2011). The Arctic states have adhered to international law, and voices predicting an Arctic scramble have also been noticeably fewer. One of the probably most vocal authors foreseeing an "Arctic meltdown," IR-scholar Scott Borgerson later reviewed his prediction, saying: "[a] funny thing happened on the way to Arctic anarchy. [...] Proving the pessimists wrong, the Arctic countries have given up on saber rattling and engaged in various impressive feats of cooperation" (2013:79). Similarly, an international project led by the Norwegian Institute for Defence Studies, GeoPolitics, in the High North 2008–2012, reached the same conclusion: no race to the North in sight. Despite strong interests, carefulness and cooperation – rather than political drama – characterizes the Arctic. (Udgaard, 2013).

Whereas things such as the global energy market and a low oil price has lowered the risk of conflicts by cooling down interests (see Keil, 2014), other things also have contributed, such as joint interests, Arctic kinship, and a symbolically important reference to the region as standing above turmoil. We "look forward to a long-term future of peace and stability in the region," the eight Ministers for Foreign Affairs declared in a joint statement (2016), celebrating the 20th anniversary of the Arctic Council. The Arctic states seem anchored to cooperation, not conflict, at least for now.

Stories told on ice-melting as unavoidably leading toward resource rushes have turned out to be nothing else than stories, narratives, versions of reality rather than *the* reality (cf. Dittmer et al., 2011; Avango, Nilsson and Robert, 2014). This shed light on the "performative aspects of geopolitics," highlighting its connection to political choices and the political salience therethrough given (Eklund and van der Watt, 2017:99). However, what is worth remembering is that stories can also shape reality. This again is worth keeping in mind due to the region's geopolitical situated-ness between to major powers whose relations, in an international context, more recently has deteriorated. Predictions of a conflict between foremost the

U.S and Russia to spillover, or at least for the potential of a spillover, into the Arctic affairs have been made (Huebert, 2016; Rahbek-Clemmensen, 2017). Indeed, the Arctic future is not pre-given but is one where different actors will fight over getting to determine the outcome (Wormbs and Sörlin, 2013). And if the future can be fought over, it turns the future into something actively – not accidentally – decided upon.

### …But Well a Small Drama of Energy-Environment Intertwinement

The 'small' Arctic drama involves two different stories to be told. One is the story about Arctic oil and gas as resources that *will* be extracted;[5] technological innovations, equipment and vessels designed for icy conditions, and floating oil rigs with a design strong enough to escape or withstand floating ice are some safety measured developed to enter what often is referred to as the 'last resource frontier.' For instance, the International Association of Oil and Gas Producers (IOGP) confirms the Arctic as "a top advocacy focus area" (2017), and similarly, the Norwegian energy company Equinor (former Statoil) introduces their activities in the offshore Arctic, saying "[w]e recognize that there is opposition to exploring the Barents Sea, but we believe it is safe and responsible to do so" (2019). As a matter of fact, from an energy political perspective, Arctic oil and gas does not only bring profit to companies, revenues to states, and possibly social and economic development, but also contributes with important and – especially when comparing natural gas with coal – cleaner energy in a world where the demand for energy seem to be on a constant rise.

The second Arctic story is about environmental pollution, like that of oil spills and short-lived climate pollutants, both expressed under the bigger heading of climate change. Extraction of fossil energy resources continues in the Arctic, despite the environmental risk such extraction (and the transportation thereof) does pose in terms of oil spills and methane leakages, in addition to the greenhouse gases it causes when being combusted. With both stories – and rather, realities – having prominent positions within Arctic cooperation, the small Arctic drama does not entail conflicts involving military aggression but rather, conflicts in interests.

Arctic warming mitigation and reaping the benefits of various commercial opportunities are interests that appear dichotomous: warmer temperatures means less ice, which means more commercial activities, which leads to even less ice, all occurring in an ongoing feedback loop. For instance, the Arctic Athabaskan Council (AAC), the Inuit Circumpolar Association, and Greenpeace together with Nature and Youth have all sued their respective governments (Canada, the U.S., and Norway), for not doing enough to slow the Arctic warming (Earthjustice, 2013; Greenpeace, 2016a; ICC Canada, 2005; George, 2006). The lawsuit against the Norwegian government was explicitly directed to the decision to allow for new and expanded oil drilling in the Barents Sea, where the legal argument put forward was such that expanded drilling would violate Norway's commitments under the Paris agreement, as well as future generations' right to safe and healthy environment (Greenpeace, 2016a).

However, environmental protection and commercial activities are *also* two interest spheres that stand in a *symbiotic* relationship to one another, where both are legitimized under the concept of sustainable development. For states, it thus becomes a question of finding an accurate balance between the three pillars of economic, social, and environmental development. An important dimension to settle concerns whose sustainable development one is trying to promote. Those who are living close to Arctic nature might not share perspective on what is considered a sustainable use of Arctic resources, and they also find themselves further away from the political decision-making and economic power. The inclusion of indigenous associations as PPs in the Arctic Council is an attempt to secure the local perspective to be consulted by the member states (*Ottawa Declaration*, 1996, Art. 2). Still, with ensured welfare, now and into the future, interpreted as the goal of sustainable development, the extraction of oil and gas for now has an important function to fill should economic and social development be promoted. As such there is a probable possibility of environmental protection being given a subordinated position when states attempt to find the balance equalizing sustainable development.

Now, *if* interest associated with energy politics contra environmental politics would be intertwined only in a one-directional manner, departing from energy, then Arctic environmental protection could not do much else than wait for the day that fossil fuel becomes dated. But if, on the contrary, it is understood as a two-directional intertwinement, then nuances could be added that have less to do with strategic thinking and materialism, and more to do with norms. From this perspective, the small Arctic drama is ultimately a drama resting upon values, and values do, indeed, change. Oil companies like Shell, for instance, caused headlines by its decision to withdraw from the Alaskan Chukchi Sea, after investing billions of dollars and years of work. In the Canadian Beaufort Sea, other oil companies have done the same[6] (Dawson, 2015; Macalister, 2015). And although it was a decision that President Trump later came to reverse, the former Obama administration in the U.S. as well as in Canada took the political decision not to extend or issue new Arctic offshore drilling leases (Goldenberg, 2015; U.S.-Canada Joint Arctic Leaders' Statement, 2016). This development is contrary to what many expected when USGS firstly published its Arctic resource assessment. To claim the Arctic rich in resources did not automatically trigger a commercial and political activity in line with that. Arctic environmental protection and cooperation could therefore, potentially, be in a better position than previously expected.

### The Arctic – A Debated 'Common'

The regulation of Arctic issues – be it environmental protection, exploration of oil and gas reserves, or control over shipping passages – is not all connected to the same governance system: some are attached to state jurisdiction, whereas other issues relate to GEG. Thus, to what extent can the Arctic states 'claim' the region – as a region belonging to them? And how is Arctic cooperation motivated by the Arctic states; is it carried out to protect state sovereignty, to avoid a 'tragedy of the common,' or perhaps both?

It has been suggested that as Arctic ice-melting continues, the Arctic Council could wish to develop its role into the future in three different ways: toward a society of sovereign states, giving priority to 'government'; toward acting like steward for the region, prioritizing environmental 'governance'; or toward being a regional security actor (Wilson, 2016). At least for now, these are dimensions all found within AC cooperation, driven by an Arctic state motivation aiming for two things: to *control* the region by, for instance, strengthening state sovereignty in relation to external actors, and to *protect* the region by, for instance, encouraging the world to globally take responsibility for the Arctic environment.

## *The Logic of Cooperation: To Enhance Regional Control*

To start with the former, from a legal perspective Arctic is not much of a global common. All land is territorially linked to the Arctic states, and most of the Artic water as well. The Arctic Ocean and related waters are clearly regulated through the UN Convention on the Law of the Sea (UNCLOS). Each coastal state – Canada, Denmark, Norway, Russia, and U.S. – has here the sovereign right to regulate foreign shipping within twelve nautical miles from its coast. The area between the coastline and 200 nautical miles further is the Exclusive Economic Zone (EEZ), which grants the state sole rights for fishing and resources. Outside this area, the state that can prove to the UN Commission on Delimitations of Continental Shelves that its continental shelf continues will be given continued right to seabed and resource extraction (UNCLOS, Art. 76). Another dimension of the Arctic as a region governed by sovereigns is found in relation to the 'Arctic exception' as listed in Article 234. Here, coastal states are given enhanced pollution prevention powers in icy waters, despite this being beyond twelve nautical miles from the coast.

However, as pointed out by Michael Byers (2013:165), with more ice turning into water, more vessels from abroad will sail the Arctic waters. At the same time, Article 234 may lose its significance, should the requirement of ice no longer be fulfilled. As such, the Arctic states could potentially lose the right to claim tougher pollution regulation, while still being the closest to any accident and the consequences therefrom. Therefore, from the perspective of regional and sovereign control, there are strong incentives for Arctic states to cooperate: AC agreements on cooperation such as *Aeronautical and Maritime Search and Rescue in the Arctic* (2011) and *Marine Oil Pollution Preparedness and Response in the Arctic* (2013) are, for instance, establishing the rules of procedure should accidents occur at the sea, sorting out issues of responsibility as well as formally establishing a willingness to help each other out in a region where ports are at long distances. It has also been important to stand united in international negotiations, for instance, on International Maritime Organization's (IMO) Polar Code, which enter into force in 2017, after almost two decades of negotiations. This code acknowledges that polar eco systems, and to some extent also Arctic communities, are vulnerable to ship operations, and thus strengthens the requirements for risk mitigation, for instance, through ice-classified ship design and training of the personnel, for those vessels that want to sail – what is described as – "inhospitable" polar waters.

Littoral states are also worried of what will happen with those water columns that are outside of their EEZs. Illustrative, in 2015, the coastal states agreed in a *Declaration concerning the Prevention of Unregulated High Seas Fishing in the Central Arctic*, not to enter the high seas to harvest any living marine resources until science provided a sign of clearance. It has been suggested that the coastal states – the so called A5 – by such an initiative tried to protect their fishing stocks from great catches made by foreign-flagged vessels, by sending a message to the rest of the world – fishing companies and nations – to follow their example and abstain from fishing (*The New York Times*, 2015). As such it would be a message on 'appropriate behavior' in the Arctic, since legally, anyone has a right to fish in this part of the Arctic Ocean. But above all, the A5 fishing agreement sends the message of the Arctic states as in control, signaling responsibility, stewardship, and perhaps, Arctic ownership. Thus, the Arctic (littoral) states' wish for resources, control, and security turns Arctic waters into something that should be considered as little 'common' as possible, especially if one with 'common' stretches on to the international society.

### The Logic of Arctic Cooperation: Promoting the Arctic as a Global Common

Since coastal states are entitled to claim continental shelves as territorial extensions, there is only a small part of Arctic seabed resources to consider as 'common heritage of mankind' (UNCLOS, Art. 136; Byers, 2013). However, from an environmental legal perspective, a regime such as 'common concern of humankind' has evolved. It is quite interpretative in terms of what to count as 'common concern' (Bartenstein, 2015:9), but climate change and biodiversity do here both qualify; although much of the seabed and its resources are connected to national territory, things such as atmosphere, water columns, and surface/land can be considered when assessing what to view as common concerns. As mentioned in Chapter 1, there is a strong ecological and climatological linkage scientifically confirmed between the Arctic and the world where a warming Arctic provides for climate feedback, i.e., causes additional climate change. Arctic warming is thus of global concern.

Still, since no single state 'owns' the environmental problem, the lack of ownership renders the Arctic region to, at least, potentially suffer from a management deficit of collective resources, a problem known as 'the tragedy of the commons' (Hardin, 1968). The tragedy refers to a perceived inability of humanity to handle common natural resources in a sustainable manner, instead causing resource depletion and overexploitation. To harvest and get access to a common, without regulations in place or clear divisions of responsibilities attached, will have the individualistic and short-term perspective to triumph what from a long-term perspective would be best for mankind and nature alike. Consequently, without cooperation to guide its development and with only unilateral actions, the Arctic faces the risk of being over exploited by states who fail to deliver in relation to the norm of environmental responsibility. However, a tragedy is an outcome that should not be considered a predestined necessity (cf. Ostrom, 1990), and a cooperative logic would thus be to have it escaped. The Arctic states *do* address responsibility – regional as well as global – in relation to environmental pollution, by referring to a 'commonality': to 'common Arctic issues,' to 'common concerns,' and to 'common objectives' (see Bartenstein, 2015; National Arctic strategies; Arctic Council declarations).

Now, to cooperate to control the region and to protect the region requires somewhat different things from actors. The more 'commonness' one is after, the more environmental responsibility it takes, and the more shared understanding one should wish interaction to bring about. Because, to achieve enhanced environmental protection and escape the tragedy of the commons has, as described by McCay (1996:117), less to do with inherited characteristics of human nature and more to do with norms and values. Rather than a tragedy, Arctic environmental pollution should therefore best be understood as a *drama* of the commons (cf. Dietz et al., 2002). Such a drama fits well with the before-mentioned 'Arctic paradox,' where no clear ending is in sight.

### The Development of Arctic Council Cooperation: An Interplay between Science and Politics

When Arctic development is on the agenda do politicians, or others, not infrequently return to the Norwegian explorer Fridtjof Nansen for symbolic guidance. Reportedly, he once said: "[t]he great thing in human life is not so much where we stand as in what direction we are moving." In March 1895, Nansen found himself standing on the Arctic sea ice. For the last one year and a half, his ship *Fram* had been stuck in the ice, all according to the initial plan since pack ice – Nansen knew – is always in constant movement. However, as it had turned out, the sea ice moved in the wrong direction: rather than drifting towards the North Pole where the achievement was looming of becoming the first man ever, the ice drifted elsewhere. It would take three more years of drifting, from north of the Laptev sea to Svalbard, until the ship eventually could break free and sail back home to Tromsø, with all men healthy and sound but with the mission yet incomplete.[7]

Just like Nansen's expedition, cooperation is about movement. In the context of Arctic environmental governance, the best-case scenario would involve a common deployment of norms on new environmental standards, in addition to increased common understanding between cooperating parties. Less successful cases would instead reveal cooperation to deploy a poor common understanding, with an interaction numb on norms and with a movement going off course. Should it be *directional* movement, however, where Arctic state interaction means less drifting and more of firmly 'keeping the compass' at environmental protection, it would be driven by learning.

Environmental science has the potential to take out the direction and spur change. It 'articulates persuasive accounts of the global nature of biological systems,' which could undermine state-based political orders by creating new narratives on norms (Miller, 2004:47). Often, scientists are transnational in their actor-ness, which means that actors such as experts no longer have to mirror preferences of state government, but could, together with norms and institutions, provide for change within nation-states (ibid.; Haas and Haas, 1995:257). The founding statute of the Arctic Council also acknowledges the role of both traditional knowledge and science and describes them to be stakeholders that contribute "to the collective understanding of the circumpolar Arctic" (*Ottawa Declaration*, 1996). These stakeholders and specialists are thus exercising – implicit or explicit – soft power. AMAP, one of the WGs, describes its mission to be precisely this, to inform policy-makers and deliver sound and science-based information. Since it often is the WGs themselves that initiate and push the research frontiers forward, they have in a sense an 'institutional life' of their own, producing knowledge in a cumulative manner. As explained by Haas (1992), anyone in control of knowledge also has access to power *over ideas*, i.e., of change.

However, the internalization of environmental norms does not happen in a straightforward way; instead it reveals both progress and setbacks. A clear example is the UN climate regime, which for three decades have searched a consensus among states, strong enough to eventually arrive at fossil fuel redundancy. This process, as others, entails a struggle between forces calling for change and forces that attempt to resist (these changes), unfolding science and politics as representing somewhat competing interests, as parallel processes of problem-solving (Jasanoff, 2004:21). Often, it is not a clear line of division between scientific findings and political decisions: political understanding is, to its nature, different from scientific understanding, since the former is composed of many more parameters – like costs, political strategy, national opinion – than what scientific facts are composed of. In that sense, to decide a political follow-up on a scientific finding means to make a political *interpretation* of that very same knowledge that science produces.

In fact, any scientific inquiry itself could be as value-free and objective that it wishes for, but its relevance will still have to go through a political process. Illustratively thereof, the SAOs were presented with a new report on climate change from AMAP WG, and responded by firmly reminding the WG of its responsibility "to interact with SAOs in such a way that SAOs will be *comfortable* [emphasis added] with developed WG recommendations" (SAO meeting Ilulissat, 2010). In environmental cooperation, in order for a scientific inquiry to gain also political relevance, scientific findings are thus tested as being justified and true, by giving way to political reason and argumentation (Habermas, 2005:222–223). A scientific assessment would consequently be likely to influence policy not just based upon the extent to which it is viewed as scientifically reliable, but to the extent it is understood to be relevant to the specific context and judged as politically acceptable (Clark, Mitchell and Cash, 2006).[8] Thus, science and politics are *together* making out the setting through which actors try to make sense of their world (Jasanoff, 2004). In the Arctic Council, state actors' consideration of the above-mentioned parameters has led to cooperative progress – the embracement of 'change' – in some, whereas cooperative developments have been restrained in others.

*Arctic Council Development: Progression*

During the years that have passed since the Arctic Council first was established in 1996, a political mission has been visible amongst the Arctic states; they not only wish to cover a larger portion of environmental problems and sustainable development issues, they also seek to have this cooperation socially *evolved*, where Arctic states are brought closer to one another. At one Ministerial meeting after another, its members have been highlighting that cooperation is getting better and is moving forward: it has gone through a widening, by the covering of more environmental issues, the acceptance of more observer members, and the creation of several free-standing and issue-specific AC offspring.[9] It has also gone through a deepening – for instance, since 2011, a secretariat has been operated from the Norwegian city Tromsø, and its members have furthermore agreed to sign binding agreements. Over the years, AC development has created an institution that (gently) starts to demand of states to follow up on the AC work – its recommendations and its agreements – and to mirror it within the national frameworks on Arctic policies and regulations.

The deepening of the AC institutional structure also has had more of a social dimensions, inclined political stability, and trust. For instance, the AC has been described to have "matured into enhanced stability," ready to accept and benefit from more observer members[10] (AC Ministerial statement by Iceland, 2015). Two key strengths of the AC have furthermore been described to be the shared sense of purpose and the shared sense of trust amongst the members (AC Ministerial statement by U.S., 2015).

Attempts have been made to increase the possibilities for common understanding to evolve, by having politics escaped as representing a divisive force. As an example, in relation to the presentation and publication of the Arctic Climate Impact Assessment (ACIA), several of the climate active Nordic countries pressed for having science and policy recommendations separated (SAO meeting Fairbanks, 2000). Through such standpoint, these states confirmed science to contain the 'good arguments' for action, whereas 'policy' rather was understood to constrain the same. Actually, the ACIA revealed how a state like the U.S., who initially was very skeptical toward climate policy recommendations and who wanted to have meetings behind closed doors, had pressure from others – specific states, PPs, and scientific community – to reach an agreement (Nilsson, 2007:139–140).

Through the ACIA and the major assessment it represents, international knowledge on Arctic climate change took a leap forward, and it is generally considered to be one of the ACs greatest achievements. Still, David Stone, the former chairman of the AMAP and highly involved in the process, described several years later how many of those working with ACIA still felt disappointment over the Arctic Council response, and the final document allowed for less political substance than hoped for; for instance, the ministers made several decisions regarding the ACIA, including policy recommendations on mitigation, adaptation, and research, but they carefully avoided implying any commitments on the GHG reductions (Stone, 2015:207–208). Once more, a common understanding on environmental politics was revealed constrained by political realities.

*Arctic Council Development: Restrained Cooperation*

Arctic Council cooperation cannot free itself from a decision-making structure built around consensus; only a few binding commitments have been made, and state sovereignty continues to have a preeminent status. To "share experiences" and "exchange best practices" are generally the way forward, as put in SAO meeting minutes and declarations, and anything touching upon sovereignty has been kept away from the negotiation table. Sensitive issues have been dealt with on an informal basis, at dinners and meetings outside the AC (Nilsson, 2007:136). Thus, AC meetings are in a way 'orchestrated spectacles' (Geertz, 1980), staging consensus-reaching in such a way that they are laid up for success by leaving tougher issues aside – at least those who are 'broadcasted an audience.'

Resistance could be noticed in hands-on things, such as a flaw in implementing scientific WGs' recommendations. The WGs are highly productive, and they have produced extensively over the years, in what should be considered an even increasing scale. However, a problem identified by its members, states observers, and PPs alike is that many reports are passing by unnoticed, while their recommendations lack requirements on follow-up. "Sadly, many thorough and good documents produced in the Arctic Council just end up on the shelf, missing out to reach both relevant stakeholder and political implementation," one observer representative explained. In the same vein, World Wildlife Fund's (WWF) Global Arctic Program adds on, "the only entities that truly take on WG recommendations 'are…Arctic Council working groups'" (Dubois and Tesar, 2014).

Implementation of the AC recommendations is furthermore complicated by these often being formulated in a vague and general way. A joint auditing of the Arctic states' participation in the Arctic Council, conducted by Denmark, Norway, Russia, Sweden, and the U.S., identified a major challenge ahead of the AC regarding how it should ensure effective national implementation of the agreed AC recommendations (Multilateral Audit, 2015). The flawed follow-up that exists within the AC has thus created calls for an upgrade of the AC structure, with a focus on implementation (Dubois et al., 2016).

But resistance is also more invisible. To produce knowledge requires decisions of bringing it about: it requires the decision to order *a* report (and not another); it requires the decision to render to financial funds available for one specific research area (and not another); and it requires the 'AC stamps of approval' and attention at Ministerial meetings. As an interviewee speaking from within the AC WGs explains, "*nothing* will be released with the AC logo attached if not previously approved by the SAO" (interview 2014). Moreover, "at the Ministerial meeting, the short amount of time – only two hours – turns working groups presentations of new reports and scientific findings into a competition for time" (ibid.). Time, one would assume here, equalizes visibility, and is a vital prerequisite if new norms should be made available to policy-makers, and eventually recognized or even internalized as of value. Following Finnemore (2014), the task is to create channels between the episteme and the decision-makers, strong enough to have 'the world to hang together' through norm transfer and shared norms, rendering states unable to resist new norms out of old habit.

## Arctic Council Evolvement: Three Time Periods

The Arctic Council has since its establishment in 1996 made imprint on Arctic development. The development of the Arctic Council is a response to stimuli of a various kind, including things such as scientific findings, economic considerations, political ambitions, senses of trust alongside threats of conflicts, in addition to a growing awareness of the Arctic environment and its interlinkage to climate weather systems. Furthermore, the development of core Arctic interests such as environmental protection and resource harvesting both require the presence of peace, stability, and cooperation to be achieved, and are in that sense strategically coupled to each other. However, such coupling takes different expressions depending on the time-period and the actors involved. Arctic relations and cooperation could be divided into three time periods,[11] and next these periods will be discussed to represent 'AC movement,' and the therein evolvement of Arctic cooperation

### Period I – 1996–2006: 'Science as Regional Attention Maker'

In 1996, Arctic cooperation was formalized into the Arctic Council. The spreading of knowledge, awareness, and cooperative (normative) models manifests the first period of AC cooperation, as primarily norm-producing and policy-shaping (Brigham et al., 2016; Pedersen, 2012; Rottem, 2015). In the beginning, there was little interest in carrying out political activities between member states, and its loose institutional structure could best be described as a high-level forum. Budget deficits and a lack of transparency and communication limited cooperative ambitions further. Arctic cooperation was marked by a 'study and talk' mentality, with little movement from words to action (Koivurova and VanderZwaag, 2007:191).

Evidence of the Arctic region and AC cooperation as of low-political importance is found if looking at the signatories of the *Ottawa Declaration* upon its founding in 1996. Apart from the host country Canada who had been very active in establishing the AC, Foreign Ministers were at this point in time conspicuous by their absence, and instead the states sent diplomats at lower levels to attend the meeting.[12] The lack of dedicated Arctic priority also took an illustrative expression prior to the next Ministerial meeting, the second. The SAOs suggested that the topics for the ministers to confer about must be of political interests "in order to get them [ministers] to come" (SAO meeting Fairbanks, 2000). A publicity stunt was suggested by Norway, linking the Ministerial meeting to some other reporting from Alaska: the Ministerial meeting itself would not be enough as an event to attract the attention from at least the Norwegian journalists, Norway said (ibid.).

In terms of specific issue areas, this period had initially one major concern: biodiversity and how it was affected by pollution. Arctic research revealed high levels of toxins and heavy metals, often originating from lower latitudes but carried northward by air or ocean currents (AMAP, 1997). Not long thereafter, the epistemic communities within the AC also asked for a green light on what they considered to be a 'key topic': to monitor and assess climate change and UV-B in the Arctic. From their perspective, climate change was viewed as "an overwhelming threat to Arctic biodiversity" (SAO meeting Anchorage, 1999), and the AMAP, CAFF, and the International Arctic Science Committee (IASC) started to carry out the ACIA. The states, though, were not that overly convinced, asking for clearer and more detailed information, not least regarding budget (ibid.).

A few years later, all doubts were gone. The ACIA was now congratulated for its success, and the SAOs expressed their appreciation to all member states having actively participated in the work (SAO meeting Rovaniemi, 2001). Indeed, the ACIA was described by the Arctic Ambassadors as "the most comprehensive work ever undertaken in the Arctic Council, and the most difficult" (SAO meeting Reykjavik, 2004). The main message of this flagship report, involving hundreds of scientists and later merged into IPCC's fourth assessment report was that of the Arctic being a climatic hub inextricably linked to the rest of the world. At this stage of cooperation, such insight did, however, not rise demands for any measures to be taken by the Arctic states themselves.

It was not just awareness – through science – that the AC tried to spread across the globe; it was also awareness of AC as a cooperative model. In time for the World Summit on Sustainable Development in 2002, the AC organized a side event presenting the AC as a model for partnership between the government and indigenous people, for the rest of the world to be inspired and learn from. It was later reported that indigenous people from many parts of the world had recognized the success of Arctic cooperation (SAO meeting Oulu, 2002; Inari, 2002). The normative dimension of the AC thus stretched beyond environmental issues per se, to constructing the Arctic region as centered on soft law and normative awareness within a range of issues, in a global context of still being a region fairly unrecognized. Although still with some reluctance, cooperation would soon take on more of a political character.

### Period II – 2007–2013: 'Political Cooperation'

The next phase of the AC cooperation did not have so much of an inside-out approach any longer – where AC states aimed for international visibility – but rather received strong impetus from the international sphere. The Norwegian ambition during its chairmanship from 2006 to 2009 was to make the AC more "potent and politically relevant," as the region gained importance in several ways (Pedersen, 2012:151). In 2007, IPCC released its fourth assessment report on global warming, confirming its anthropogenic origin. The same year, the International Polar Year started, gathering researchers and students alike in hundreds of research projects aiming for climate awareness. Suggestively, the Arctic sea ice of that year was plummeting, measuring the lowest levels of ice in September since measurements began in 1979 (National Snow & Ice Data Center, 2017).

In the AC, one tuned up the discourse on climate change into a question of Arctic destiny. At the Ministerial meeting in Tromsø in 2009, illustratively Iceland as a first victim of the global financial crisis had fallen deep and hard just the year before, put it into perspective: "[...] the changes we are now witnessing in the Arctic as a result of human induced climate change are of much more profound and long lasting nature than our economic difficulties which we will surely overcome in a few years" (AC Ministerial statement by Iceland, 2009). Members started to demand state actions on mitigation, which was introduced through short-lived climate pollutants, although it would take six more years until (in 2015) any kind of agreement was reached on the matter.

To play with words such as the 'Arctic is hot' (in a dual sense) during this period made a political, economic, and environmental/climatological sense.[13] Applications to become an observer increased, in parallel to the idea of the Arctic as increasingly accessible. In a rat race, the Arctic states started to present national Arctic strategy documents, which – if not before – now placed the Arctic on their respective foreign policy agenda (Rottem, 2015:52); Norway was first to go in 2006, Russia followed in 2008, Canada and the U.S. presented their national strategies in 2009, Finland in 2010, and Denmark, Iceland, and Sweden all did so in 2011. The view on the Arctic as a potential gold mine grew strong, confirmed by the USGS (2008), where the Arctic was estimated to house huge hydrocarbon reserves. For many of the Arctic states – as well as the global oil and gas market – this meant good news. In 2008, the annual average price for a barrel of oil was record high at 99 U.S. dollars/barrel, and although it dramatically fell the following year, it was again up on even higher levels between 2011 and 2014, peaking at an annual average of close to 112 U.S. dollars (U.S. Energy Information Administration (EIA, 2020).

Against this context, Russia created international headlines in 2007 by planting a titanium flag on the seabed, 4,261 meters beneath the northern most point of the world. Canada was quick to condemn the flag planting, which they saw as a legally unauthorized attempt for land grabbing (*Independent Online*, 2007). However, the Russian explanation of a scientific investigation in parallel to a domestic election soon coming up seemed reasonable enough to many. The phrase "[t]he Arctic is ours" (as the expedition leader Chilingarov reportedly had said) was not interpreted by most people as a literal attempt to actually claim the north (Byers, 2013:92). Instead, during this period, an expectation on politics spelled out, where diplomacy, cooperation, and adherence to international law prevailed, further exemplified through a comment made by Denmark's former Foreign Minister Per Stig Møller. In a leaked cable between him and a U.S. diplomatic official, he referred to the U.S. as not yet having ratified the UNCLOS, 'joking': "if you stay out, then the rest of us will have more to carve up in the Arctic" (Jones and Watts, 2011). It was a joke that created attention, interpreted by some as nonetheless containing grains of truth (see Dodds, 2011).[14]

Within the AC, one kept calm. "Arctic is not in a legal vacuum," Jonas Gahr Støre declared, rejecting any such thing as a 'race for the Arctic,' and pointed toward responsible and modern states governed under regulatory schemes such as UNCLOS, IMO rules, and various environmental and fishery regimes (AC Ministerial statement by Norway, 2009; 2011). However, at the same time, differentiating politics found its way into the AC, when the five Arctic coastal states – the so called A5 – on firstly Denmark's initiative and then on Canada's, met in 2008 and 2010 to discuss Arctic 'go abouts' without inviting the non-littoral states or the PPs. These meetings were met with criticism of being noninclusive. Reportedly, in a speech at the closed session of the 2010 meeting, Hillary Clinton voiced support for participation by everyone with legitimate interests and hoped the Arctic would showcase an ability to work together rather than creating new divisions (Sheridan, 2010). However, these meetings resulted in agreements among the coastal states to commit to international law (UNCLOS), where the need for such meetings perhaps above all were tellingly of a region gone politicized. In 2011, and from then on, the U.S. would also choose to send Secretaries of State to Ministerial meetings, adding political weight thereof.

Given the new context, this second phase of the AC development was focused on *preparation*: to prepare the region for increased activity to come. Organizationally, this meant that one was pre-occupied with increasing the efficiency of the organization as an operating entity, and having it mature. With more activity expected in the region, the states agreed to the first binding agreement ever in the history of the AC, on marine Search and Rescue (SAR) (2011), with the primary aim of enhancing maritime coordination and cooperation amongst the Arctic states. An intensified focus on oil pollution soon led to the second binding agreement, now on oil pollution preparedness and response (2013). Although these agreements were weak, they – together with the work, for instance, on black carbon – revealed an Arctic region that becomes more regulated. In that sense, the AC now started to move from merely policy-shaping into policy-making, with policy recommendations of practical significance attached, and where politics and science in that sense become more intertwined.

### Period III – 2014–Present: 'Cooperation under Insecurity'

*Arctic Insecurity*

If the second period was about preparing the Arctic for the change caused by more human activity, the third period represented change as an insecurity. This is partly expressed through the economy, where Arctic resources might not be as lucrative after all, causing the Arctic standard image to be somewhat challenged (Granholm, 2016). Whereas the oil price was close to 113 U.S. dollars per barrel in 2013, it drastically began to fall in the autumn of 2014 reaching low levels just slightly above 30 U.S. dollars per barrel in January 2016. Although it quickly again rose, the oil price has at large remained at levels of about 50–70 U.S. dollars barrel (EIA, 2020). Yet, although extraction costs differ significantly throughout the offshore Arctic, even this price level is considered too low in order to boost the investments in the Arctic hydrocarbon extraction.[15]

In the Arctic, Russia's ongoing militarization is furthermore a cause of concern. However, whereas this at times aggressive militarization has offended the other Arctic states' air-space or internal waters, much of this military development could be linked to the Russian wish to control and protect its strategic assets and infrastructure as its Arctic shoreline is getting more exposed due to Arctic warming and increased shipping (Pezard, et. al., 2017). In line with what would be Russia's Arctic ambition, the Northern Sea Route has become more trafficked; in 2013, barely four million tons of cargo was transported along the Northern Sea Route, which in 2018 had grown to 18 million tons of cargo (Humpert, 2019a).

Even though this is only a fraction of all cargo transported along the traditional route via the Suez Canal, it is expected to continue to grow. Much of the cargo transported on the NSR is oil, coal, and not least liquefied natural gas from the Russian Yamal peninsula. By 2030, Russia plans to increase NSR traffic by 90 millions of tons (Middleton, 2020). Since marine conditions remain tough with less sea ice, new and expanded marine traffic in the Arctic, shipping as well as tourism raises the need for best practices, cooperation, and common guidelines on safety and security.

Outside of the Arctic, tensions among the Arctic states have also been sensed in relation to Russia and Ukraine, the annexation of Crimea, and interventions in the Middle East. To the U.S., the Arctic behavior of Russia – as well as China – represents an increased level of conflicts in the region (U.S. Department of Defense, 2019). China's presence in the Arctic is growing – for instance, in relation to research, shipping, and investments. China describes itself to be a near-Arctic power who takes an interest in the Arctic due to climate change. But China also has mining deals with Greenland, and likely as a result of this – attempting to have the Chinese presence reduced in the Arctic – the U.S., through President Trump, announced a desire to buy Greenland. However, it was a business offer described by the Danish Prime Minister as being absurd, and which had Denmark to put Greenland the highest on its national security agenda (Peter, 2019).

Thus, global political tensions are noticed in the Arctic, where they have actually found their way into the AC, although in the shape of environmental concerns, at the last Ministerial meeting, the U.S. Secretary of State brought foreign policy up, explaining how China's environmental destructive behavior in other regions should be of common concern to all the Arctic states (AC Ministerial statement by the U.S., 2019). But as regards insecurities pertaining to Arctic environmental *cooperation* per se, the U.S. has here added on, most notably by withdrawing from the Paris agreement, which has a ripple effect on Arctic warming cooperation as well. To an organization like the AC, whose backbone rests upon environmental protection and to – from there – ensure cooperation and peace, recent developments within the environmental field spur an insecurity regarding where the region is headed at.

### Arctic Council Cooperation Ahead

Despite insecurities, what then would support a continuous Arctic Council cooperation? Well, quite a few things. In relation to Russia and its international military behavior, the AC states clinger to the idea of an Arctic peace zone, ring-fenced the rest of the world. Sweden – similar to others – has formulated it as follows: "[t]hat the Arctic continues to be a zone of peace and stability is paramount. [...] there is no problem that cannot be solved through cooperative relationships" (AC Ministerial statement by Sweden, 2015). In addition, Finland explains: "[i]t is in no-one's interest to let problems elsewhere to impact negatively on Arctic cooperation and the Arctic environment" (AC Ministerial statement by Finland, 2015). Actually, a joint U.S.-Russia proposal on routing measures and precautionary areas in the Bering Sea and the Bering Strait has been argued by the two states to demonstrate their continued cooperation in the Arctic (SAO meeting Levi, 2018). The WWF was thanked for encouraging them to develop such a proposal, which serves to increase maritime security and the protection of wildlife habitat (ibid.).

The strategy chosen by members has been to separate the 'external world' from the 'Arctic world,' with the latter assumed to run by its very own logic. To maintain cooperation and to keep up environmental protection is – if not in all aspects, then at least in quite a few – the predominant guiding principle for any (commercial) activity in the Arctic. Perhaps Russia best formulates this logic:

> Russia sees huge potential in the Arctic Council to promote and expand a constructive agenda for our common region, built on the basis of national interests of all Arctic states [...] There is no room for confrontation or aggravation of tension in the Arctic region.
> (AC Ministerial statement by Russia, 2015)

But Arctic Council is not only about material interests and mutual interdependence, but also about unique Arctic cooperation described to be a "success story," revealing the Arctic states play by the same rules (AC Ministerial statement by Denmark, 2015). Perhaps, under current political insecurities, what will provide for 'success' in future cooperation is described by the U.S. who urges each council member to ensure that "bonds of trust and responsibility remain unbroken" (AC Ministerial statement by the U.S., 2019). As such, what predominantly seem to threaten Arctic cooperation under political insecurity would be lowered levels of trust among the Arctic states.

Still, so far, the general trend of Arctic development is greater regulation – outside the AC as well as within. As part of its strategic work, the Arctic Council – its efficiency and effectiveness – is sought to be strengthened (AC Rovaniemi Ministerial statements, 2019). In the Arctic Council, furthermore, stakes have been raised by discussing more 'demanding' issues such as oil pollution prevention and reduction of black carbon. A third binding agreement has also entered into force, the Agreement on Enhancing International Arctic Scientific Cooperation (2017). In periods of political insecurity, one seems once again to cling to science as the unifier and guiding light. "Science will lead the pathway forward," Sweden's foreign minister has reassured; Norway has called for a robust AC capable of finding "common solutions to common challenges," and Iceland has explained that the clear mandate of the AC and its focus on things such as regional knowledge-building has allowed cooperation to continue, putting aside global political tensions (AC Ministerial statement by Sweden,

2017; Norway, 2017; Iceland, 2019). Following this line of thought, the Arctic would no more foremost be about opportunities and challenges, but solutions. Even more so, common solutions often require science, and they always require cooperation.

## Conclusions: On Course for Arctic Cooperation?

Cooperation within the Arctic Council could be described as a movement with direction, toward policy-making: it has deepened in terms of institutional structure, with a standing secretariat and where calls for follow-up and implementation have intensified. After twenty years of effort by the PPs to get access to a funding mechanism that would help them to more fully participate in the AC decision-making, a charitable fund has now been established, which is expected to increase the capacity of the PPs AC participation considerably (Gamble, 2016). The Arctic Council has also expanded, for instance, oil emergency preparedness has been broadened and specified to include oil spill prevention, and previous spreading of Arctic warming as a global threat has been complemented by demands on reductions of black carbon by Arctic states themselves. Approaches to nature, as in the need for adapting (due to environmental risks) has been transformed into a resilience approach, signaling a shift in value where nature is looked upon as capable of recovering only that much.

Then, the AC cooperation is a cooperation that also seems to drift, with an identified problem of lacking requirements for implementation and follow-up mechanisms attached to political decision-making (Multilateral audit, 2015). The development of the AC cooperation could be seen driven by a correlation between science and politics, but where politics still would have the final say and could slow the cooperative pace down. Yet, also from a strictly political and material interest-perspective, deepened Arctic cooperation makes sense: on the one hand, for national interests and on the other, for signaling 'Arctic control' to the outside world. A5 meetings, where legal supremacy is given to international law to govern Arctic state relations, jurisdictional claims, and fishing quotas, all portray the coastal states as taking control over the Arctic, thus, making external involvement redundant.[16]

Whereas order in the Arctic state relations and order in the Arctic governance in many ways is a prerequisite for revealing the Arctic (Ocean) as less of a common and more of a region protected by sovereignty, science has played a big part in the AC development by providing for enhanced knowledge and understanding of the Arctic environment. In parallel, states have also come to manifest their region and relations as prominently peaceful and cooperative. Thus, although current development may bring forth some cooperative hardships, AC cooperation stays committed a 'movement with direction,' revealing few things as indicative of Arctic relations soon turning into discord rather than cooperation.

# Notes

1 The Arctic Circle is the definition used to identify the eight member states of the Arctic Circle. However, the definition is somewhat narrow, not least from a climatic perspective. Other ways of defining the Arctic is through either the July 10 isotherm, or based on the tree line: the Arctic border is where the forests end and the northern tundra begins. Compared to the Arctic Circle, both of these definitions move further south in relation to the Bering Sea, and also further south in Canada and below Greenland (AMAP, 1997).

2 For example, in 1878, the Finland-Swede Adolf Erik Nordenskiöld sailed the Northeast Passage; in 1905, the Norwegian Roald Amundsen and his expedition became the first to sail the Northwest Passage, and in 1909 (although disputed) the American Robert Peary together with four Inuit men set their feet on the North Pole.

3 For instance, each of the years between 2014 and 2019 exceeded the temperature since measurement began around 1900, where the temperatures of 2016 and 2018 during winter were 6 degrees Celsius above 1981–2010 average. Since 1979, multiyear sea ice thickness has declined by 90 percent (IPCC, 2019).

4 Although the U.S. was not innocent itself, they saw the radioactive pollution as a threat resulting from past Soviet activities, and – overall – was not that easily convinced of the benefits of Arctic cooperation (Dingman, 2015:2). However, it did not take long until Arctic pollution was recognized as even greater in magnitude and diversified concerning emission sources. People are contaminated foremost through traditional food (living) from the oceans: whales, seals, some fish species, and polar bears. Despite an ongoing global effort to reduce persistent organic pollutants, new contaminants are continuously transported to the Arctic (see, for instance, the AMAP report *Human Health in the Arctic*, 2015).

5 This message has echoed with strength at industrial O&G conferences. For instance, the Arctic Oil and Gas Conference, 2014; Arctic Technology Conference, 2015; and IEA Gas and Oil Technology, 2017.

6 For instance, BP, Exxon Mobile, and Chevron.

7 Nansen himself left his ship to traverse the white landscape by feet. By the time the ship had sailed back home to Tromsø, Nansen also returned from his expedition and met up with his crew in the city, pluming of being the farthest north any man previously had been at a latitude of 86,14 degrees north, although not reaching the whole way to the northern most point on earth. As a comfort to Nansen perhaps, his specially constructed ice-strengthened ship would some years later be the very same ship that took his fellow compatriot Roald Amundsen to the South Pole, as the first man ever.

8 These variables could be translated into Habermas' vocabulary about validity claims (1984).

9 These AC offspring do operate freestanding from the AC, but they have the same eight Arctic states as members, and are: the Arctic Coast Guard Forum (ACGF), the Arctic Economic Council (AEC), and the Arctic Offshore Regulators Forum (AORF). Voices have been raised whether this sort of issue delegation, which serves to separate Arctic issues from the table of the AC, really is desirable, or if it rather weakens the Arctic Council's mission and mandate.

10 All members do not share this Icelandic position on observers. Both Canada and Russia were presumed to oppose an expansion of new observer members, and have therefore only reluctantly accepted such an expansion of observers

(Jakobson and Lee, 2016:114). More recently, the U.S. Senior Arctic Official also expressed concerns over further expansion of observer members, rhetorically asking whether current organizational structure effectively could cope with considering up to 40 observers' opinions (Balton, 2017).

11  These periods are divided based on the combined sum of foremost information gained from SAO meeting minutes and secondary literature; interviews too have helped to shed light on cooperative climate and characteristics as varying over time.

12  The informing state here seems to be the U.S., who declared that they would not send any such foreign minister, thereby causing Russia to also send a representative with lower diplomatic weight; the deputy foreign minister rather than the foreign minister himself. The Scandinavian countries also chose to send officials from other ministries than the foreign affairs (Pedersen, 2012:149).

13  See, for instance, former SAO G. Lind, in *The Economist*, 2012.

14  Another interpretation would instead be that it was a comment signaling trust and common understanding, where the U.S. delaying of ratification should be dealt with through cooperation.

15  An exemption would be Norway, whose operating industry in the Norwegian part of the Barents Sea calculates with a break-even price of 40 dollars per barrel. It could be compared to extraction in the Russian part of the Barents Sea, where conditions are less favorable with harsher weather and more ice. To extract here is normally considered to require a price of about 90–100 dollars per barrel in order to be profitable (IEA Oil and Gas and Technology Conference, 2017).

16  The wish to self-govern the Arctic could help explain the reluctance by Norway, Denmark, and Iceland to allow the OSPAR Commission – Protecting and Conserving the North-East Atlantic and Its Resources – to create a marine protected area in the Arctic Ocean (Greenpeace, 2016b). These states argued such discussions to be designated to the Arctic Council, and no other (international) forum (SAO meeting Fairbanks, 2016).

## References

AC Ministerial Statement by Denmark. (2015). *Ministerial meeting in the Arctic Council – Intervention by the Danish Foreign Minister.* 9th Ministerial Meeting. Iqaluit, Canada, April 24, 2015. (All Ministerial statements accessed via oaarchive.arctic-council.org)

AC Ministerial Statement by Finland. (2015). *Statement by Mr. Erkki Tuomioja, Minister for Foreign Affairs of Finland.* 9th Ministerial Meeting. Iqaluit, Canada, April 24, 2015.

AC Ministerial Statement by Iceland. (2019). *Speech by the Minister of Foreign Affairs of Iceland, Mr. Gudlaugur Thór Thórdarson.* 11th Ministerial Meeting. Rovaniemi, Finland, May 7, 2019.

———— (2015). *Statement by Iceland.* 9th Ministerial Meeting. Iqaluit, Canada, April 24, 2015.

———— (2009). *Address by H.E. Ásta Ragnheiður Jóhannesdóttir, Minister of Social Affairs of Iceland.* Ministerial Roundtable on Climate Change: Opportunities and Challenges Created by a More Accessible Arctic. Tromsø, Norway, April 29, 2009.

AC Ministerial Statement by Norway. (2017). *Statement by Mr. Børge Brende, Norway's Minister of Foreign Affairs.* 10th Ministerial Meeting. Fairbanks, Alaska, May 11, 2017.

———— (2011). *Inlegg på Arktisk Råds møte på Grønland.* 7th Ministerial Meeting. Nuuk, Greenland, May 12, 2011.

———— (2009). *Norwegian MFA Jonas Gahr Støres welcoming remarks at the sixth Ministerial Meeting of the Arctic Council.* Tromsø, Norway, April 29, 2009.

AC Ministerial Statement by Russia. (2015). *Statement by Mr. S.E. Donskoy at the Arctic Council Ministerial Meeting.* 9th Ministerial Meeting. Iqaluit, Canada, April 24, 2015.

AC Ministerial Statement by Sweden. (2017). *Speech by Minister for Foreign Affairs Margot Wallström.* 10th Ministerial Meeting. Fairbanks, Alaska, May 11, 2017.

———— (2015). *Statement by Sweden.* 9th Ministerial Meeting. Iqaluit, Canada, April 24, 2015.

AC Ministerial Statement by the U.S. (2019) *Statement by Secretary of State Michael R. Pompeo.* 11th Ministerial Meeting. Rovaniemi, Finland, May 7, 2019.

AC Rovaniemi Ministerial Statements. (2019). Joint Declarations of the Foreign Ministers of the Arctic States at the 11th Ministerial Meeting of the Arctic Council, Held in Rovaniemi, Finland, 7 May, pp. 1–22.

Arctic Council. (2004). *ACIA – Impacts of a Warming Arctic. Arctic Climate Impact Assessment.* ACIA Overview Report. Cambridge: Cambridge University Press, pp. 1–140.

Arctic Environmental Protection Strategy (AEPS). (1991). *Declaration on the Protection of the Arctic Environment.* June 14, Rovaniemi, Finland.

Arctic Monitoring and Assessment Programme (AMAP). (2017). *Snow, Water, Ice and Permafrost in the Arctic. Summary for Policy-Makers.* Oslo, Norway.

———— (2015). *AMAP Assessment 2015: Human Health in the Arctic.* Oslo, Norway.

———— (1997). *Arctic Pollution Issues: The State of the Environment Report 1997.* Oslo, Norway.

Avango, D., Nilsson, A. E. and Robert, P. (2013). "Assessing Arctic Futures: Voices, Resources and Governance," *The Polar Journal,* Vol. 3:2, pp. 431–446.

Balton, D. (2017). *Panel Discussion: The Arctic in a Global Context.* At the Arctic Frontiers Conference, Tromsø, Norway. January 24, 2017, published online January 25, 2017. Accessed from www.youtube.com/watch?v=AWx2yzWm77Y.

Bartenstein, K. (2015). "The 'Common Arctic': Legal Analysis of Arctic and Non-Arctic Political Discourses," in Heininen, L., Exner-Pirot, H. and Plouffe, J. (eds.). *Arctic Yearbook 2015. Arctic Governance and Governing.* Iceland: Northern Research Forum.

Borgerson, S. G. (2013). "The Coming Arctic Boom: As the Ice Melts, the Region Heats Up," *Foreign Affairs,* Vol. 92:4, pp. 76–89.

———— (2008). "Arctic Meltdown: The Economic and Security Implications of Global Warming," *Foreign Affairs,* Vol. 87:2, pp. 63–77.

Brigham, L., Exner-Pirot, H., Heininen, L. and Plouffe, J. (2016). "Introduction – The Arctic Council: Twenty Years of Policy-Shaping," in Heininen, L., Exner-Pirot, H. and Plouffe, J. (eds.). *Arctic Yearbook 2016. The Arctic Council: Twenty Years of Regional Cooperation and Policy-Shaping.* Iceland: Northern Research Forum.

Byers, M. (2013). *International Law and the Arctic. Cambridge Studies in International law and Comparative Law.* Cambridge: Cambridge University Press.

———— (2009). *Who Owns the Arctic? Understanding Sovereignty Disputes in the North.* Vancouver/Toronto/Berkeley: Douglas & McIntyre.

Chater, A. (2016). "Explaining Russia's Relationship with the Arctic Council," *International Organizations Research Journal,* Vol. 11:4, pp. 41–54.

Clark, W. C., Mitchell, R. B. and Cash, D. W. (2006). "Evaluating the Influence of Global Environmental Assessments," in Mitchell, R. B., Clark, W. C., Cash, D. W. and Dickson, N. M (eds.). *Global Environmental Assessments: Information and Influence.* Cambridge, MA: MIT Press.

Dawson, C. (2015). "Exxon Mobil, BP Suspend Canadian Arctic Exploratory Drilling Program in the Beaufort Sea." *The Wall Street Journal*, June 26, 2015.

Dietz, T., Dolšak, N., Ostom, E. and Stern, P. S. (2002). "The Drama of the Commons," in Ostrom, E., Dietz, T., Dolšak, N., Stern, P. C., Stonich, S. and Weber, E. U. (eds.). *The Drama of the Commons.* Committee on the Human Dimension of Global Change, National Research Council. Washington, DC: National Academy Press.

Dingman, E. (2015). "Arctic Council Environmental Initiatives: Can the United States Promote Implementation?" in Heininen, L., Exner-Pirot, H. and Plouffe, J. (eds.). *Arctic Yearbook 2015. Arctic Governance and Governing.* Iceland: Northern Research Forum.

Dittmer, J., Moisio, S., Ingram, A. and Dodds, K. (2011). "Have You Heard the One about the Disappearing Ice? Recasting Arctic Geopolitics," *Political Geography,* Vol. 30:4, pp. 202–214.

Dodds, K. and Nuttall, M. (2016). *The Scramble for the Poles.* Cambridge and Malden: Polity Press.

——— (2011). "The WikiLeaks Arctic Cables," *Polar Record,* Vol. 48:2, pp. 199–201.

Dubois, M-A and Tesar, C. (2014). "Making It Stick – A New Approach to Implementing Arctic Council Decisions and Recommendations," in Heininen, L., Exner-Pirot, H. and Plouffe, J. (eds.). *Arctic Yearbook 2014. Human Capital in the North.* Iceland: Northern Research Forum.

Dubois, M-A., Eichbaum, B., Shestakov, A., Sommerkorn, M. and Tesar, C. (2016). "Arctic Council Upgrade: WWF Arctic Programme Note," in Heininen, L., Exner-Pirot, H. and Plouffe, J. (eds.). *Arctic Yearbook 2016. The Arctic Council: Twenty Years of Regional cooperation and Policy-Shaping.* Iceland: Northern Research Forum.

Earthjustice. (2013). *Petition to the Inter-American Commission on Human Rights Seeking Relief from Violations of the Right of Arctic Athabaskan Peoples Resulting from Rapid Arctic Warming and Melting Caused by Emissions of Black Carbon by Canada.* Summary. Submitted April 23, 2013, on behalf of Arctic Athabaskan Council.

*The Economist.* (2012). "Arctic Politics. Cosy Amid the Thaw. The Arctic Council Works Well – Because of the Region's Riches." March 24, 2012.

Eklund, N. and van der Watt, L-M. (2017). "Refracting (Geo)political Choices in the Arctic," *The Polar Journal,* Vol. 7:1, pp. 86–103.

Energy Information Administration (EIA – U.S.). (2020). *Petroleum and Other Liquids. Europe Brent Sport Price FOB.* Accessed on January 17, 2020, from www.eia.gov.

Equinor. (2019). *Why It's Responsible to Explore the Barents Sea.* Accessed on December 12, 2019, from www.statoil.com/en/what-we-do/responsible-drilling-in-the-barents-sea.html.

Exner-Pirot, H. (2016). *The Arctic Factor: Can Regional Cooperation Thaw Relations between Canada and Russia.* Interview with Huebert, R. and Exner-Pirot, H. In *Open Canada,* February 17, 2016.

Finnemore, M. (2014). "Dynamics in Global Governance: Building on What We Know," *International Studies Quarterly,* Vol. 58:1, pp. 221–224.

Gamble, J. (2016). "Commentary. The Arctic Council Permanent Participants: A Giant Step Forward in Capacity & Support," in Heininen, L., Exner-Pirot, H. and Plouffe, J. (eds.). *Arctic Yearbook 2016. The Arctic Council: Twenty Years of Regional Cooperation and Policy-Shaping.* Iceland: Northern Research Forum.

Geertz, C. (1980). *Negara. The Theatre State in Nineteenth-Century Bali.* Princeton: Princeton University Press.

George, J. (2006). "ICC Climate Change Petition Rejected." *Nunatsiaq News,* December 15, 2006.

Goldenberg, S. (2015). "Obama Administration Blocks New Oil Drilling in the Arctic." *The Guardian,* October 16, 2015.

Gorbachev, M. (1987). Mikhail Gorbachev's speech in Murmansk at the ceremonial meeting on the occasion of the presentation of the order of Lenin and the Gold Star to the city of Murmansk. October 1, 1987.

Granholm, N. (2016). *Arktis under förändring – standarbilden utmanas.* Swedish Defence Research Agency, Report number: FOI-R – 4268 – SE.

Greenpeace. (2016a). *Historic Lawsuit against Arctic Oil.* Press release October 18, 2016. Accessed from www.greenpeace.org/international/en/press/releases/2016/lawsuit-arctic-oil-norway-historic/.

———— (2016b). *Norway, Denmark and Iceland Prevent Protection of Unique Arctic Area.* Press release June 24, 2016. Accessed from http://archivo-es.greenpeace.org/espana/es/news/2016/Junio/Norway-Denmark-and-Iceland-prevent-protection-of-unique-Arctic-Area-/.

Haas, E. (1992). "Introduction: Epistemic Communities and International Policy Coordination," *International Organization,* Vol. 46:1, pp. 1–35.

Haas, P. M. and Haas, E. B. (1995). "Learning to Learn: Improving International Governance," *Global Governance,* Vol. 1:3, pp. 255–284.

Habermas, J. (2005). *Truth and Justifications.* Cambridge, MA: MIT Press.

———— (1984). *The Theory of Communicative Action. Reason and the Rationalizaton of Society,* Vol. 1. Cambridge, MA: Polity Press.

Hardin, G. (1968). "The Tragedy of the Commons," *Science,* Vol. 162:3859, pp. 1243–1248.

Heininen, L. (2015). "The Arctic Region as a Space for Trans-disciplinary, Resilience and Peace," *Arctic and North,* Issue 21:4, pp. 69–73.

Hough, P. (2013). *International Politics of the Arctic: Coming in from the cold.* London and New York: Routledge.

Huebert, R. (2016). *The Arctic Factor: Can Regional Cooperation Thaw Relations between Canada and Russia?* Interview with Huebert, R. and Exner-Pirot, H. In *Open Canada,* February 17, 2016. Accessed from www.opencanada.org/features/arctic-factor-can-regional-cooperation-thaw-relations-between-canada-and-russia/.

Humpert, M. (2019a). "Russia's Northern Sea Route See Record Cargo Volume in 2018." *High North News,* February 19, 2019.

*Independent Online.* (2007). "Canada Rejects Flag-Planting as 'Just a Show." August 3, 2007.

International Association of Oil and Gas Producers (IOGP). (2017). *Arctic Environment.* Accessed on April 11, 2017, from www.iogp.org/arctic-environment/.

Interviews. (2014–2015). Interviews conducted between December 2014 and December 2015 with state representatives and other representatives active in Arctic Council cooperation and negotiation.

Inuit Circumpolar Council Canada (ICC Canada). (2005). *Inuit Petition Inter-American Commission on Human Rights to Oppose Climate Change Caused by the United States of America.* Petition of December 7, 2005.

Intergovernmental Panel on Climate Change (IPCC). (2019). "Chapter 3: Polar Regions," In: *IPCC Special Report on the Ocean and Cryosphere in a Changing Climate.* Pörtner H. O., Roberts, D. C., Masson-Delmotte, V., Zhai, P., Tignor, M., Poloczanska, E., Mintenbeck, K., Alegría, A., Nicolai, M., Okem, A., Petzold, J., Rama B. and Weyer, N. M. (eds.). In press.

Jakobson, L. and Lee, S-H. (2016). "North East Asia Eyes the Arctic," in Jakobson, L. and Melwin, N. (eds.). *The New Arctic Governance.* SIPRI Research Report No. 25. Oxford: Oxford University Press.

Jasanoff, S. (2004). "Ordering Knowledge, Ordering Society," in Jasanoff, S. (ed.). *States of Knowledge. The Co-production of Science and Social Order.* London: Routledge.

Joint Statement from the Arctic States. (2016). *The Arctic Council: A Forum for Peace and Cooperation.* A joint statement from Ministers of the Arctic States on the occasion of the Arctic Council's 20th anniversary, on September 19, 2016.

Jones, M. and Watts, S. (2011). "Wikileaks Cables Show Race to Carve Up Arctic." BBC Newsnight, May 12, 2011.

Keil, K. (2014). "The Arctic: A New Region of Conflict? The Case of Oil and Gas," *Cooperation and Conflict,* Vol. 49:2, pp. 162–190.

Keskitalo, E. C. H. (2004). *Negotiating the Arctic. The Construction of an International Region.* New York and London: Routledge.

Koivurova, T. and Molenaar, E. J. (2010). *International Governance and Regulation of the Marine Arctic.* Three reports prepared for the WWF's International Arctic Programme. Oslo: WWF International Arctic Programme.

Koivurova, T. and VanderZwaag, D. (2007). "The Arctic Council at 10 Years: Retrospect and Prospects," *University of British Columbia Law Review,* Vol. 40:1, pp. 121–194.

Macalister, T. (2015). "Shell Abandons Alaska Arctic Drilling," *The Guardian,* September 28, 2015.

McCay, B. J. (1996). "Common and Private Concerns," in Hanna, S. S., Folke, C. and Mäler, K-G. (eds.). *Rights to Nature: Ecological, Economic, Cultural and Political Principles of Institutions for the Environment.* Washington: Island Press.

Middleton, A. (2020). "Northern Sea Route: From Speculations to Reality by 2035." *High North News,* January 7, 2020.

Miller, C. A. (2004). "Climate Science and the Making of a Global Political Order," in Jasanoff, S. (ed.). *States of Knowledge. The Co-production of Science and Social Order.* London: Routledge.

Multilateral Audit. (2015). *The Arctic Council: Perspectives on a Changing Arctic, The Council's Work, and Key Challenges. A Joint Memorandum of a Multilateral Audit on the Arctic States' National Authorities' Work with the Arctic Council.* Conducted by the Supreme Audit Institutions of Denmark, Norway, the Russian Federation, Sweden and the United States of America, May 5, 2015.

National Snow and Ice Data Center (NSIDC). (2017). *Arctic Sea Ice News and Analysis.* Accessed from http://nsidc.org/arcticseaicenews/

*The New York Times.* (2015). "'No Fishing' at the North Pole." Editorial. July 21, 2015.

Nicol, H. N. and Heininen, L. (2014). "Human Security, the Arctic Council and Climate Change: Competition or Co-existence?" *Polar Record,* Vol. 50:1, pp. 80–85.

Nilsson, A. E. (2012). "Knowing the Arctic: The Arctic Council as a Cognitive Forerunner," in Axworthy T. S., Koivurova T. and Hasanat W. (eds.). *The Arctic Council: Its Place in the Future of Arctic Governance.* Collection of papers originally presented during a conference organized by Munk-Gordon Arctic Security Program, University of Lapland.

——— (2007). "A Changing Arctic Climate. Science and Policy in the Arctic Climate Impact Assessment." Dissertation. Linköping Studies in Art and Science No 386. Department of Water and Environmental Studies, Linköping University. Linköping: UniTryck.

Nord, D. C. (2016a). *The Arctic Council: Governance within the Far North.* New York: Routledge.

Ostrom, E. (1990). *Governing the Commons.* Cambridge: Cambridge University Press.

*Ottawa Declaration.* (1996). *Declaration on the Establishment of the Arctic Council.* Signed by the Arctic States on September 19, 1996.

Pedersen, T. (2012). "Debates over the Role of the Arctic Council," *Ocean Development & International Law,* Vol. 43:2, pp. 146–156.

Peter, L. (2019). "Danes see Greenland Security Risk Amid Arctic Tensions." *BBC News,* November 29, 2019.

Pezard, S. Tingstad, A. Van Abel, K. and Stephenson, S. (2017). *Maintaining Arctic Cooperation with Russia: Planning for Regional Change in the Far North.* Santa Monica, CA: Rand Corporation.

Posner, E. (2007). "The New Race for the Arctic." *The Wall Street Journal,* August 3, 2007.

Rahbek-Clemmensen, J. (2017). "The Ukraine Crisis Moves North. Is Arctic Conflict Spill-Over Driven by Material Interests?" *Polar Record,* Vol. 53:1, pp. 1–15.

Rottem, S. V. (2015). "A Note on the Arctic Council Agreements," *Ocean Development and International Law,* Vol. 46:1, pp. 50–59.

Runge Olesen, M. (2015). "Common and Competing Interests," in Jokela, J. (ed.). *Arctic Security Matters.* ISS – EU Institute for Security Studies. Report No. 24, June 2015, pp. 1–82.

SAO Meeting Minutes. (2018). *Arctic Council Senior Arctic Officials Meeting Levi, Finland.* March, 22–23, 2017. Report, pp. 1–19. Accessed via oaarchive.arctic-council.org

——— (2010). *Ilulissat, Greenland.* April 28–29, 2010. Final report, pp. 1–23.

——— (2004). *Reykjavik, Iceland.* November 22–23, 2004. Draft minutes, pp. 1–13.

——— (2002). *Oulu, Finland.* May 15–16, 2002. Minutes, pp. 1–22.

——— (2001). *Rovaniemi, Finland.* June 12–13, 2001. Minutes, pp. 1–13.

——— (2000). *Fairbanks, Alaska, U.S.A.* April 27–28, 2000. Minutes (revised – 10/03/00), pp. 1–37.

——— (1999). *Anchorage, Alaska, U.S.A.* May 5–6, 1999. Minutes (revised – 28/09/99), pp. 1–17.

Sheridan, M-B. (2010). "Clinton Rebukes Canada at Arctic Meeting." *The Washington Post,* March 30, 2010.

Stone, D. P. (2015). *The Changing Arctic Environment. The Arctic Messenger.* Cambridge: Cambridge University Press.

Udgaard, N. M. (2013). "Kappløpet mot nord er avlyst." *Aftenposten*, March 21, 2013.

*United Nations Conventions on the Law of the Sea – UNCLOS* (1982). UN General Assembly, December 10, 1982.

U.S.-Canada Joint Arctic Leaders' Statement. (2016). *The White House: Office of the Press Secretary.* December 20, 2016. Accessed from https://obamawhitehouse. archives.gov.

U.S. Department of Defense. (2019). *U.S. Department of Defense Arctic Strategy.* Report to Congress, June 2019.

U.S. Geological Survey. (2008). *Circum-Arctic Resource Appraisal: Estimates of Undiscovered Oil and Gas North of the Arctic Circle.* U.S. Department of the Interior. USGS Fact Sheet 2008–3049.

Wilson, P. (2016). "Society, Steward or Security Actor? Three Visions of the Arctic Council," *Cooperation and Conflict,* Vol. 51:1, pp. 55–74.

Wormbs, N. and Sörlin, S. (2013). "Assessing Arctic Futures. Voices, Resources, Governance," in *Mistra Arctic Futures in a Global Context.* Final Report 2011–2013.

Young, O. R. (2011). "The Future of the Arctic: Cauldron of Conflict or Zone of Peace," *International Affairs,* Vol. 87:1, pp. 185–193.

# 4 Role-Playing in the Arctic Council

## Introduction – The 'Theatrical Act' of Cooperating

*Theater state* describes states as dramaturgically governed, where power in international relations is less an issue of coercion and more of convincing ideas. Ideas, in such a setting, are not unobservable mental stuff but made visible in symbols, spectacles, and culture (Geertz, 1980:35). Although theater state is a concept originally and specifically applied to the nineteenth century's Bali, it still captures something symbolically relevant for present-day environmental diplomacy. Today, international meetings represent partly orchestrated spectacles on certain social realities, for instance, on how environmental problems should be understood (Campbell et al., 2014). The climate summit that manages to send the message of political awareness and determination – that political elites have risen to the challenge and have dedicated solutions – paves way for the summit to be considered a success (Death, 2011:7).

In the Arctic Council, every second year, the Ministerial meeting sends messages of peace and unity amongst its members, accompanied by assurances of a dedicated protection of the Arctic region and its environment. For instance, Sweden – at that particular time acting as chair – made the following call: "Let the message from all of us here in Kiruna be loud and clear. We are committed to do *whatever we can* [emphasis added] to protect the fragile Arctic environment" (AC Ministerial statement by Sweden, 2013). It is a powerful speech-act, but the promise it epitomizes is yet constrained: protecting the environment by doing whatever one can may take one far, or it may take one nowhere. This chapter aims to increase the understanding of what the states engaged in Arctic cooperation really 'can' do, given that states in international relations are role-players following scripts. More concretely, the chapter provides for a descriptive analysis of what type of actor each Arctic state is when it comes to the Arctic in general and the Arctic Council specifically; what are their Arctic state of affairs like, how do the Arctic states see themselves as actors within an Arctic context, and what would their behavior be like when entering the cooperative arena of the AC? Apart from shedding light on the social dimension of Arctic relations, the chapter serves as a background to the following analytical chapters, by mapping the roles of the Arctic states.[1]

### Behavioral Patterns and the Expression of Roles

The analysis is conducted in relation to the SAOs level of the AC, which is somewhat different in character compared to Ministerial meetings: instead of spreading messages directed at an audience globally and at home, thereby, possibly, acting theatrical, SAO meetings are concerned with the operational activities of the AC. As institutionalized meetings, SAO meeting cannot be reduced to agents writing strategic manuscripts alone but must also consider prescriptions from, for instance, scientific WGs, PPs (indigenous associations), and fellow state members. Representing an interactive arena, SAO meetings may reveal an interaction effect in the state behavior, or role performance, being envisioned.

The roles of the Arctic states are partly painted through material interests and national preconditions, and partly through structural and social components derived from prescriptions and other forms of behavioral expectations. The role mapping relies on publicly available meeting minutes from SAO meetings, held for a period of twenty years.[2] About two meetings a year have been held, where each and every meeting generally has resulted in a report of about 10–25 pages describing the meeting scene; it provides information on displayed state characteristics by bringing up comments, initiatives, and concerns put forward by the individual SAOs (i.e., member states).

Each country analysis is conducted according to the following structure. Firstly, with help from foremost national Arctic strategy documents a role conception is developed, which is elaborated upon through the state's national interests and ideas in and of the Arctic, and of Arctic cooperation. Secondly, a role performance is presented, which equalizes the state behavior witnessed in the Arctic Council. The behavior here is conceptually as well as, to some extent, practically, different from role conceptions, in such a way that it is exercised under the potential impact of socialization – using the lingua of George Herbert Mead – transforming ego into *I* and alter into *Me*. Following M. Finnemore (1996), who in Chapter 2 was referenced as saying that states are not always aware of their wants, here, in relation to their performances, states may be told of them. Thus, role performances highlight how roles are turned into social positions, relative to others. Lastly, a short comparison between role conception and role performance reveals the interaction effect, i.e., it reveals interaction to be a socialization process, possibly grinding the edges off when it comes to national role conceptions and gives rise to a role (better) fitted for the social context.

The reader should not understand each role description as a comprehensive resume of all the things states have done or said during the AC history. Issue-wise, what in this chapter gets discussed is partly decided in relation to Chapters 5 and 6, to add as much background information as possible. By accounting for small details related to behavior, rather than major projects brought ashore, this chapter follows the same idea as the rest of the book, namely, to highlight the general through the specific, the official through the personal, the state through the human.

*Mapping the Roles – Some Initial Findings*

A mapping of states' role performances in the Arctic Council reveals all eight states acting as cooperators, all speaking vibrantly about the environment – the Arctic Ocean, land, and atmosphere – as well as the right of the indigenous populations to a secure and traditional way of living. Those five states that are recognized as coastal states – Canada, Denmark, Norway, Russia, and the U.S. – are placing issues of sovereignty high, and they approach the Arctic in more unilateral manner, whereas the three non-coastal states are acting more like followers, especially in issues stretching beyond their territorial borders. The four biggest states, either in terms of Arctic interest and activities or area-wise – Canada, Norway, Russia, and the U.S. – are the most visible in meeting minutes, in terms of comments and initiatives.[3] Still, with environmental protection being such a large part of cooperation, 'smaller' states also have a say, in an organization that members describe to build on trust and where a good discussion climate is characterizing meetings (see AC Ministerial statements; interviews 2014–2015). A role therefore is a more complex construction than based only upon material power, territorial sizes, and access to resources.

The role mapping reveals the following roles: Canada – Protector, Denmark – Pragmatic (coastal state) kid-brother, Finland – Reserved team player, Iceland – Follower with ambitions, Norway – Restless know-how leader, Russia – Responsive informant, Sweden – Teacher on demand, and U.S. – Innate leader. These roles are forged after each country's cooperative characteristics and capture therefore conceptions and behavior as role *patterns*. Next, a description follows.

## Canada

*Arctic interest: to preserve the national identity by protecting the North.* To Canada, the Arctic is fundamental to its national identity. The self-perception is that of being 'people from the North,' no matter whether living on the southernmost borders or up in the cold (Huebert, 2011). An explicit goal of the Arctic policy would therefore be to exercise strong sovereignty while also ensuring Northerners an influence over their own destiny by improving and devolving northern governance (Arctic strategy, 2009:foreword; 9; 28). As a result of the latter, Canada released a new Arctic policy in 2019, codeveloped with northerners and indigenous governments to close the gap between Ottawa and the Canadian Arctic. It is explained to put "the future into the hands of the people who live there," and focuses on issues such as education, infrastructure, and economic development (Arctic policy, 2019).

In order to have its people safe, secure, and defended, Canada seeks to demonstrate its sovereignty, for instance, in the shape of strengthened emergency management (ibid.). With Canada measuring the second biggest state area-wise, and with the second longest coastline in the Arctic,[4] there would be geopolitical reasons for this. The Northwest passage, the shipping line from the Pacific Ocean to the Atlantic, goes through the Canadian archipelago. The rapid ice-melting can soon turn previously un-navigable pack-ice into open water, exposing the coast of Canada to frequent passages and increasing the need for surveillance and border security, amongst others. However, Canada has since the 1970s combined an exercise of sovereignty with a protective stewardship role as well (Burke, 2017), protecting marine ecosystems, wildlife, and biodiversity alike. Illustratively, although Canadian waters are estimated to have oil and gas deposits worth billions, oil and gas drilling leases here have been made off limits (U.S.-Canada Joint Arctic Leader's Statement, 2016).

Few countries are understood as severely affected by climate change as Canada, who therefore is calling for urgent action on climate mitigation (Arctic strategy, 2009:8; Arctic policy, 2019). Still, over the years, Canada's international climate behavior has not held a steady course on the matter; after being a leading actor during the Kyoto negotiations, it pulled out in 2011 only five years later to ratify the Paris agreement. Nationally, the government has been perceived as backlogging on climate measures[5] (Earthjustice, 2013; Lunn, 2015). Whit Justin Trudeau as Prime Minister, Canada has claimed to be back as an international climate leader. In practice, however, a discrepancy is revealed between an international Canadian proactiveness and a national implementation, and Canada has a low rating when it comes to climate change performance and mitigative action-taking (Abedi, 2019; CCPI, 2020).

*Interest of Arctic cooperation: to protect sovereignty and promote indigenous rights.* The Canadian role conception is that of being a responsible actor – someone who promotes cooperation, diplomacy, and international law (Arctic strategy, 2009). However, sometimes this conception has collided with its emphasis on sovereignty; in international settings Canada has acted unilaterally to protect the interests of the indigenous population, for instance, by not being part of the International Whaling Commission, by not having polar bears enlisted as endangered species, or by being opposed to the EU as an AC observer due to the former's seal trade ban (EUR-Lex, 2009; CBC News, 2009). Together with Russia, it has contributed to the rhetoric on the Arctic resource race (Nicol and Heininen, 2014). To Canada, such a race requires increased territorial control, due to the historic context of all past Canadian threats to have come via the Arctic.[6] Canada is increasingly stressing the need for a strengthened rule-based international order in the Arctic, which could bolster its international leadership while ensuring protection of national interests (Arctic policy, 2019).

Nationally, Canada has, over the years, had a tough Arctic rhetoric on, for instance, increased military spending, the Arctic as 'ours,' and

Canadian Arctic as being "non-negotiable" (CBC News, 2010; Hilde, 2014:150). This should, however, most likely be directed toward a home audience. Because, even in turbulent times of, for instance, Russian flag-planting, a leaked cable uncovered that the Arctic Council indeed is the primary vehicle of solving Arctic issues (Barrera, 2011). To Canada, the AC continues to be the pre-eminent forum for cooperation, and the state wants to strengthen its capacity to contribute to the AC's work (Arctic and Northern Policy Framework, 2019). The AC's firm commitment to science and evidence-based decision-making has been described as something to be grateful for, since it has enhanced understanding of the North, and turned the world into better shape (Government of Canada, 2016). As such, the Canadian role conception seems to struggle in uniting its national strife for enhanced sovereignty with Arctic prescriptions on multilateralism, causing disturbances in how to present itself.

### *Canada's Role Performance in the Arctic Council: Protector*

Contrary to most other states, Canadian Ministerial statements have in the past not been published on the AC website, with a few exceptions. This yet entails useful signals on state behavior, and correlates well with Canada's role performance at the SAO level where Canada acts like a protector; indeed, acting as a protector, it would be redundant to spread national remarks, visions, and concerns beyond the cohorts invited. Two things are protected; sovereignty – translated into integrity at the more personal level – and the interests of indigenous communities.

*Theme: sustainable development.* When Canada chaired the AC during 2013–2015, Leona Aglukkaq – an Inuk politician from the Nunavut territory – led it through a program called Development for the People of the North, which was designed to showcase the north and let the voices of the Northerners be heard. "We eat what we hunt and that is not going to change," ICC once said in the beginning of the AC's history (AC Ministerial meeting Barrow, 2000). This message, similar to the PPs in general, has been taken up by Canada, who considers itself to have valuable experience to offer concerning the strengthening of indigenous people in their capacity to manage their own affairs (cf. AC Ministerial meeting Barrow, 2000; SAO meeting Inari, 2002; Haparanda, 2012; Whitehorse, 2013). To Canada, there would be no Arctic North without the people inhabiting it, and there would be no Canada without the North. Relationships with the Indigenous people of Canada have been declared to be "an absolute priority" of the Canadian government (AC Ministerial statement by Canada, 2017).

Even though sustainable development is emphasized over climate change, the latter has still been considered a threat since the outset, when Inuit hunters reportedly had to start with sunscreen, indicating something was wrong (AC Ministerial meeting Barrow, 2000; SAO meeting Fairbanks, 2000). Although Canada *outside* of the AC withdrew from the Kyoto protocol, it stayed committed within the AC context; together with

Sweden and the U.S. they, for instance, launched the Climate and Clean Air Coalition to reduce short-lived climate pollutants, they offered to co-lead the AC task force on black carbon and methane, and they were pleased to report on rather recent emission reductions regarding black carbon (SAO meeting Stockholm, 2012; Haparanda, 2012; Rovaniemi, 2018).

*Characteristics as an AC partner.* Canada is a confident actor that is keen to initiate things and to take on leadership. It is also keen to voice its opinion: for instance, it *firmly* encourages, as well as *requires*, fellow members to refer to the Arctic in UN negotiations (SAO meeting Haparanda, 2012). It stands out as a significant outreacher that wants to spread the *Arctic message* onward to the world: they organize show case events, host COP meetings 'to put global focus on Arctic climate issues,' and produce documentaries about unique Arctic relations between states and indigenous peoples (SAO meeting Khanty-Mansiysk, 2005; Yakutsk, 2005; Yellowknife, 2014a; 2014b). If the Arctic, i.e., Canadian, conditions should be improved and protected, it takes an international audience to listen, Canada argues.

In AC, Canada is amongst the leaders who take on projects, not being fearful about informing the other members of its achievements: rationality rather than feelings should guide behavior, Canada seems to reason. Canada is also an actor that stands out as favoring justified and reputable behavior. For instance, when Sweden criticized Russia for having too strict legal regimes for anyone being able to do research in the Russian Arctic, Canada responded by clarifying such obstacles as existing not only in Russia but also in other states (SAO meeting Khanty-Mansiysk, 2005). Thus, Canada is sensitive toward injustices and wants to promote "good values," and often acts as a motivator trying to encourage everyone else to follow (Nord, 2016a; 2016b). Although not overly appraised by the PPs, Canada is – compared to the other Arctic states – standing quite well off.

### What Social Interaction Has Brought About

With the positive exception of short-lived climate pollutants, where climate is a stronger part of the Canadian role performance than what its national climate commitment at first would indicate, Canada is performing a role that is close to what issue-wise is prescribed by the Arctic strategy; 'economy' – understood as a capacity builder – and sovereignty are both high on the agenda. The AC role is one of protector, but whereas the role conception portrays a rather 'tough' protector of the North, Canada's role performance in the AC turns this role more into a 'soft' ditto. The Canadian role conception struggles in juggling the different meanings attached to being a 'protector,' which is revealed in tensions between rhetoric and practice: rhetorically, there has been an element of orchestrated spectacles on tough sovereignty, but in the AC setting a reputable behavior built around gratitude and encouragement seems closer at heart as cooperative guiding principles. As such, social interaction pushes Canada in a multilateral direction, where it is evident that a protection of the North – and in parallel the Canadian identity – requires the help from others.

## Denmark

*National Arctic interest: to protect the environment while creating development.* Access to the Arctic is, in the case of the Kingdom of Denmark, granted through Greenland. Since 1953, when Greenland's status changed from a colony to a constituency of the Kingdom of Denmark, Greenland has sought greater autonomy.[7] A not unimportant but rather possible trajectory to such autonomy runs via increased national economic development. Resource extraction is here an important component, where there are, for instance, promising oil and gas reserves found in Greenlandic offshore sediments. Denmark's Arctic strategy (2011) centers around peacefulness and safety, and where self-sustaining growth and development can thrive under close environmental supervision and protection. Focus is not on state security but on societal security, which includes issues such as the support of the indigenous populations of Greenland, marine shipping, and oil and gas development. The strategy explains particularly climate change as key (ibid.:50). Indeed, Denmark follows the same pattern as the rest of the Nordic states in its focus on the environment, and in a global context it repeatedly is in the lead concerning climate change performance (CCPI, 2020).

*National interest of Arctic cooperation: to be strengthened as a global Arctic player.* To strengthen its status as a global player in the Arctic is an explicit objective of Denmark (Arctic strategy, 2011:11). Because of Greenland, Denmark escapes being "just another small state" (Offerdal, 2014:81), and instead gets to sit at the same table as big nations like Canada, Russia, and the U.S., where it "punches above its weight" (Jacobsen, 2016). Denmark is keen on acting as a sovereign Arctic nation, and surveillance and territorial claims – on an area surrounding the North Pole and on the Hans Island[8] – become important tools thereof. The Joint Arctic Command is patrolling the coast of Greenland and Faroe Islands to, for instance, ensure marine pollution prevention, SAR, and fishing vessel inspection (Defense Command Denmark, 2017).

Regardless of how important sovereignty may be, Denmark wishes for enhanced Arctic cooperation. Since Denmark, in relative terms, describes itself to be an Arctic great power who expects an upgrade to its Arctic strength as the Arctic has become more politically unstable (Taksøe-Jensen, 2016), Denmark's Arctic behavior could indeed be interpreted as moving away from cooperation (Rahbek-Clemmensen, 2016). However, it is a state that pays close attention to what signals its action will send, and Denmark has been keen to eradicate any misunderstandings that might lead to escalatory measures by others; to upgrade strength is connected to the wish to defend Arctic cooperation (Fogh Rasmussen, 2015). Similarly, the 'Arctic 5' Ilulissat meeting organized by Denmark in 2008 was believed to have sent 'strong signals' on coastal states' commitment for a responsible development of the Arctic (Arctic strategy, 2011:14), just as high-level political participation would 'send a strong signal' on peaceful Arctic cooperation, when celebrating the tenth anniversary of the meeting (SAO meeting Levi, 2018). Flag-planting episodes on Hans Island – in 2002, 2003, and 2004 – have furthermore been interpreted as representing a staged performance on power, broadcasted to the home audience close to elections (Byers, 2013:14).

The role conception of Denmark is one of being a global Arctic player, who is in control. Environmental protection is one area where Denmark conceives of itself to be such a player, and Arctic climate change specifically becomes a way for Denmark to pursue and exercise an 'active climate diplomacy.' With international cooperation as well as sovereignty being key to Arctic development, Denmark seems preoccupied with the cause of clarity to eradicate misunderstandings that otherwise could arise from its unilateral behavior.

### Denmark's Role Performance in the Arctic Council: Pragmatic (Coastal State) Kid-Brother

In the AC, Denmark's role performance is pragmatic: rather than a departure in *what could be,* Denmark departs from *what is.* And for the people living in the Arctic, a priority would be development, which entails the developing of living as well as non-living resources, oil and gas included (AC Ministerial statement by Greenland). However, oil and gas activities cannot be disconnected from (environmental) risks, and Denmark acknowledges that 'whenever humans are involved risks are associated and cannot be fully avoided' (ibid.). Similarly, although Denmark commented on the first oil spill agreement in 2013 as an agreement that perhaps could have been more ambitious, it was also remembered that "sometimes perfect can be the enemy of good" (AC Ministerial statement by Denmark, 2013). Such reasoning represents well Denmark in the AC as a conscientious actor who is not carried away with big words on improvement. Reality rather than visions would be the guiding principle for Denmark. However, whereas its role conception is telling of a global Arctic leader, its performance struggles to be a naturally 'big' state; in this coastal state context, Denmark performs like a pragmatic kid-brother.

*Melting ice.* In the AC, Denmark is pushing for regional and global actions on climate change. As a single issue, it has resulted in more comments from Denmark than, for instance, marine safety and oil and gas. With Greenland housing one of the biggest ice sheets in the world, second after the Antarctic ice sheet, Denmark has taken a particular interest in ice-melting, calling out for global action on this matter (SAO meeting Narvik, 2007; Tromsø, 2007; Ilulissat, 2010; Tórshavn, 2010). When the ACIA report was under process, Denmark hoped it would be "as extensive as a scientific report could be" (SAO meeting Oulu, 2002). In general, Denmark has attempted to reach out internationally on the issue of climate change, and it has also been willing to raise the stakes for the region itself; Denmark has, for instance, referred the mitigation of global warming to how Arctic states should respond to the ACIA findings, where they should take responsibility for their "large share of global greenhouse emissions" (AC Ministerial statement by Denmark, 2009). Denmark has also attempted to link international climate negotiations to AC Ministerial meetings (SAO meeting Whitehorse, 2015). But Denmark is also dedicated to work toward even more sustainable economic development, why it, for instance, must be ensured that AC recommendations on black carbon reduction does not hinder economic growth in Arctic communities (AC Ministerial statement by Denmark, 2019; SAO meeting Juneau, 2017).

*Characteristics as an AC partner.* The quite realistic/pragmatic approach to cooperation is reflected in rather explicit expectations, where Denmark expects others to do their part: Denmark agrees to provide financial assistance and expects at the same time others to follow (SAO meeting Anchorage, 1999); it has signed the Long-Range Transboundary Air Pollution (LRTAP) and is shocked that only Canada has followed (SAO meeting Washington, 1999); moreover, Denmark rhetorically asks why it is so difficult for some countries to make a commitment on pollution reductions, which is possible only through international cooperation (AC Ministerial meeting Barrow, 2000). This more pragmatic stance on Arctic cooperation resonates well with why Denmark only quite rarely speaks about other states: if you require something of someone else, you might be targeted by expectations and requirements in return. And a Danish role performance seems keen on not promising more than it can deliver.

Denmark is a confident actor, unconcerned toward comments made by others; for instance, Iceland's critique over being excluded from the Ilulissat A5 meeting was met with the remark, "capacity at the venue might be an issue" (SAO meeting Narvik, 2007). As a possible prolongation of its Arctic interest of enforcing sovereignty, Denmark cares for actors' integrity; meeting minutes from Denmark's chairmanship period from 2009 to 2011 reveal a note taker that to a high extent was decoding actors, thus preserving their integrity. And when Denmark organized an Arctic Information Day, the press was not invited, despite a Danish chairmanship objective of 'international outreach.' However, at that particular time, the A5 again had met separately, and a consensus was no longer as successfully obtained as in the past. A wish to protect the integrity – the imagery – of Arctic cooperation from outside critique could have impinged upon the Danish reluctance to invite the press.

To have a realistic and pragmatic perspective on cooperation, however, is not the same as not cooperating, and Denmark is generally considered as a good and trustworthy partner and a good problem solver, who, for instance, has criticized the U.S. for limiting the role of the AC (Offerdal, 2016:81). Together with Norway and Sweden, Denmark has acted to significantly strengthen the AC to have the region's best interests in mind, without losing sight of the needs of the international community (Nord, 2016b:48–49).

### What Social Interaction Has Brought About

Denmark's role performance, just like its role conception, reveals a state worried about climate change and socioeconomic conditions for its population. It views itself as an operating littoral state, with all the responsibility that comes along regarding protection of sovereignty. Performance-wise, Denmark plays this role with the confidence, realism, and pragmatism suitable for 'an Arctic power.' At the same time, in the Arctic strategy, Canada, Norway, Iceland, and U.S. are identified as close partners, sharing interests and challenges associated with the Arctic Ocean. When performing in the AC, however, it is not clear that whether these states really are supported by Denmark, or that Denmark is being supported by them. Although its role conception reveals an actor like a global player on societal security, Denmark's AC performance is more of a pragmatic (coastal state) kid-brother. Identification might be an issue for Denmark, where playing the roles of a great state with huge Arctic territory does not always comes naturally; in most other international contexts Denmark is not amongst the 'big players' and it also remains the least visible amongst the A5 nations. As such, social interaction does not really reveal – or confirm – Denmark to have the weight it foresees within its role conception. Others' expectations, imagined or experienced, are likely putting strains on a Danish role performance.

## Finland

*National Arctic interest: to provide solutions for sustainable development.* Finland is a coastal state, although not an Arctic one; rather than bordering the Arctic Ocean to the north it borders the Baltic Sea to the west. Finland, just like Sweden, lacks many of those national interests that are associated with an Arctic coastline. Nonetheless, the Arctic identity is strong, where one-third of the population lives above 60 degrees north (Arctic strategy, 2013). It considers itself to have reconciliatory competences to offer Arctic activities, capable of uniting business development and environmental sustainability. Indeed, regarding the Finnish role conception, the Arctic strategy identifies Finland as an actor that truly understands, respects, and operates within the conditions dictated by the Arctic environment. This includes an understanding of climate change, and the sustainable use of natural resources. Indeed, economic and other activities are constrained by the environment, and before entering activities, it is vital to know about environmental risks, have a preparedness, and work to have them prevented, Finland argues (ibid.:17, 38).

With implicit reference to its diversified expertise, Finland considers itself as a state with legitimacy to engage as a leader in the sustainable development of the region, with the aim of providing innovative products and practices that can overcome problems associated with Arctic development (Arctic strategy, 2013; Strategy update, 2016). It has the tools and knowledge, and further invests in research and education to "increase its appeal as a world-class expert on the Arctic" (ibid., 2013:24). Contributions are for business sectors directed, for instance, at energy efficiency and renewable energy, but Finland is above all a leading nation in marine shipping, for instance, as regards the manufacturing of icebreakers where it has one of the best capacities in the world (Bergh and Klimenko, 2016:62). Marine technology in general would be a Finnish cutting-edge sector, for instance, through winter navigation systems, offshore oil and gas drilling technologies, and technology on oil spill recovery in ice-covered waters (Arctic strategy, 2013:26–27).

*International cooperation: a means of Arctic access.* Security is of high importance to Finland, bordering as it is to Russia. However, it does not dwell upon security and sovereignty issues in the strategy (Offerdal, 2014:87), but instead Finland advances the importance of Arctic cooperation, with the Arctic Council as the primary forum which preferably can be more deepened and institutionalized (Arctic strategy, 2013:14). Furthermore, Finland is dedicated to the task of bringing the EU closer to the Arctic, by having it granted observer status in the AC. This would create a clearer and reinforced Arctic strategy on behalf of the EU, where the Arctic should be one of EU's priorities regarding international relations (Arctic strategy, 2013:15; Strategy update, 2016). It is through cooperation and international affairs that Finland exerts its national interest, since Finland itself does not naturally navigate the Arctic waters or possess any natural resources therein.

Finland has declared an overall Arctic objective to be about showcasing solution-minded Finnish leadership (Ministry for Foreign Affairs, 2016). Whereas one dimension of this leadership centers on the before-mentioned technical know-how, another relates to Finland's history of bringing attention to the region's environmental needs; and a third pertains to the importance of Arctic (higher) education. Actually, it was Finland who in 1989 invited the seven other Arctic states to the meeting later leading up to the *Rovaniemi Declaration* (i.e., AEPS); also, it was Finland who made use of its EU chairmanship in 2006 to push for a Northern Dimension perspective within the Union (AC Ministerial statement by Finland, 2006). As such, Finland conceives itself to have played a proactive role in international cooperation (Arctic strategy, 2013:8), and there is a proud history of Arctic promotion – centered on the Arctic environmental protection – within Finland.

### Finland's Role Performance in the Arctic Council: Reserved Team Player

In the AC, Finland is a reserved and low-voiced team player: to speak and inform about national activities is simply not part of the role. In the same vein, to criticize how other states approach Arctic issues or act uncooperatively is also distant to Finland, who instead is keen on acting considerately. Finland does not keep its feet on the brakes, but rather delivers in a punctual manner suitable a team player.

*Theme: protection of eco-systems.* Finland has distinguished itself as a state talking little about climate change in the SAO settings. Now, this says little for a state with more introvert characteristics, who also is talking very little about maritime security and safety, despite this being a field of big expertise. Rather than climate change specifically, Finland prefers a more inclusive approach to environmental protection, and has explained that challenges from climate change cannot be solved one by one or in isolation (AC Ministerial statement by Finland, 2017). Environmental protection, and very much so the marine life, has been the core of Finland's Arctic approach ever since the first meeting was held in Rovaniemi in 1991. To bring up Rovaniemi, as Finland occasionally has done, could represent a way for Finland to add weight to its words, to prove to be a foresighted[9] actor that understands and knows about the Arctic (see SAO meeting Oulu, 2017; AC Ministerial statements, 2013; 2015). It was also in this city Finland more recently arranged a well-received "Environmental super week," inviting all Arctic Environment Ministers to discuss Arctic environmental protection (SAO meeting Rovaniemi, 2018). According to Finland, any plans to exploit natural resources should be guided by environmental assessments, and science should, as befits conceptions of being an expert, guide decision-making and policy-drafting (AC Ministerial statements by Finland, 2002; 2013; 2015).

*Characteristics as an AC partner.* Within the reserved team player role is imbedded high integrity and respect for other members. To Finland, it does not come naturally to judge others, neither positive nor negative, which also brings forward a low frequency regarding commenting on or appraising others' behavior. However, when it happens, Canada and Russia are paid positive attention, in addition to WGs and not least the PPs. As a state, it has taken its fair share of organizing events, and performs like a reliable partner. Reliability is, for instance, spelled out when the state withdraws its offer to act as a leader on a project on Arctic biodiversity, with reference to other states as having 'more profound experience on the matter' (SAO meeting Tromsø, 2007). Thus, since 'leader' would be amongst the behavioral characteristics most frequently applied by Finland, *viewed relatively*, Finland is as a state keen on performing where they have knowledge and capacity; for instance, it has been ready to offer expertise in a newly introduced topic such as meteorology (SAO meeting, Levi, 2018; Rovaniemi, 2019; AC Ministerial statement by Finland, 2019). Apart from situations like this, the leadership role per se seems to be rather uncomfortable to wear to Finland.

### What Social Interaction Has Brought About

Finland's role conception is one of being an Arctic expert, with competence and leadership to offer to ensure Arctic sustainable development. Although Finland did apply much of such an expert role for the first ten years of AC cooperation, in addition to its recent chairmanship, Finland – just like Iceland – have become less active (or possibly less visible) in parallel to the Arctic becoming more politicized. When performing in the AC, Finland does not have its competence reflected in high vocals. Rather, Finland performs as a reserved and low-voiced team player, confident – so it seems – that the rest will search for the competence when needed. Similarly, Finland does not talk on issues pertaining to economy, such as oil and gas or marine transportation and safety, thereby avoiding encroaching upon states in their private domain: because, just as Finland respects the Arctic environment, it respects Arctic state sovereignty too. The problem-solutions offered by Finland are thus given only if asked for, and to act as an expert is only part of the role if Finland can sense that others are requesting it.

### Iceland

*National Arctic interest – to be acknowledged as an Arctic coastal state.* Iceland borders just south of the Arctic Circle and considers itself to be a fully valid coastal state due to its territorial waters reaching into the Greenland Sea (Arctic policy, 2011:6). Surrounded by water, almost everything 'Arctic' is about the sea. Equipment, technology, and instruments relating to navigation at sea are increasingly being exported, but the top export product is fish (European Commission, 2017a). With fishing being a fundamental part of Icelandic economy, history, and identity, a growing concern for climate change is the disturbance it might have on migratory fishing patterns, in addition of it being perceived as a threat to the well-being of people and communities, which should be prevented using all available means (Arctic strategy, 2011:2). Still, global warming and ice-melting are nonetheless viewed with some reassurance; a new trans-Arctic shipping route could bring economic opportunities, not least via Chinese cargos and shippers (Bergh and Klimenko, 2016; Offerdal, 2014).

Together with Russia, Iceland opposes the Norwegian Fishery Protection zone that runs around the coast of Svalbard in that it obstructs fishing beyond Norwegian fishing fleets. Iceland refers to the water as international, which would make this water fishable by quotas.[10] Iceland has also started to investigate its oil and gas assets and issued leases, but oil and gas exploration in the Icelandic continental shelf is described to still be in an early phase (National Energy Authority of Iceland, 2017).

*Interest of Arctic cooperation: to gain visibility.* A higher frequency of shipments in and close to the Icelandic EEZ increases the need for a well-functioning SAR capacity. This and illegal fishing are two areas where Iceland are in need for good cooperation. Moreover, although Iceland conceives itself to be an Arctic coastal state, it is not yet recognized as such. By pushing for recognition as a coastal state, Iceland tries to safeguard opportunities for its people. Overall, Iceland's presentation of itself as a coastal state would be directed toward the international sphere, and not least the Arctic Council cooperation which is considered the most important forum on arctic issues, to gain more influence (Arctic strategy, 2011:1; Dodds and Ingimundarsson, 2012). Iceland points toward the importance of not having member states excluded from important decisions, referring to the A5 meeting in Ilulissat 2008 and Canadian Chelsea in 2010 (Arctic strategy, 2011:6).

The Arctic strategy sets out to bring the international community to Iceland by organizing meetings and conferences, and in relation to this objective, Iceland has initiated the conference The Arctic Circle, held annually in Reykjavik. Here, the explicit goal would be to encourage dialogue and cooperation between multitudes of Arctic stakeholders. Hence, Iceland's official policy for the Arctic would be like a compound strategy aiming for Icelandic visibility.

### Iceland's Role Performance in the Arctic Council: Follower with Ambitions

Possibly due to Iceland being a small country population-wise, with no extraction of offshore energy resources and no national defense forces, Iceland struggles in its achievement of gaining the visibility and influence spoken about in the national Arctic strategy. Its role performance is telling of a low-key state that is not a big and frequent talker, who does not really initiate new projects, and who does not declare itself leader. Instead, Iceland operates quite silently, with focus kept on increased AC efficiency and good governance. Although Iceland by no means lacks ambitions, it performs the role of someone who is following rather than leading the way.

*Theme: oceans and climate.* With all passages to Iceland going by the sea, marine environment and protection are highly prioritized issues; for instance, the working group Protection of the Marine Environment (PAME) opened its secretariat in Akureyri, Iceland, in 1999, and more recently, Iceland took initiative to organize an Arctic Ocean Minister's meeting, while also announcing marine litter to be an Icelandic chairmanship priority (SAO meeting Hveragerði, 2019; Rovaniemi, 2019). That oceans are kept at the Icelandic heart, also when chairing the AC, should not come as a surprise, Iceland has explained (AC Ministerial statement by Iceland, 2019). Overall, issues such as climate change, marine environment, and maritime safety, rather than unilateral oil, gas, or fishing, have been Iceland's focus in the AC.

Just like Iceland in the beginning of AC's history strived to gain international recognition for the AC (SAO meeting Fairbanks, 2000), it now strives to get recognition for the AC *and* itself. Following its Arctic strategy, it could be assumed that Iceland tries to take up space, for instance, in relation to various high-level meetings in addition to climate change; at COP 21, Iceland offered the AC its booth in order to increase Arctic visibility in climate negotiations, and it has offered to host the side-events at future COP's (SAO meeting Anchorage, 2015; Fairbanks, 2016; Hveragerði, 2019).

*Characteristics as an AC partner.* The Icelandic role has been centered around being a communicator and an administrator that cherishes values such as transparency, order, and efficiency. After its first chairmanship, which ended in 2006, Iceland became for a period close to invisible in meeting documents. Whereas the AC at that time overall changed toward a more anonymous note-taking style, possibly as a result of a region going more politicized, the financial crisis that struck Iceland in 2008 and onward likely impacted as well. Already in 2005, Iceland was concerned over the significant value drop of the Icelandic Kroner, which impacted negatively on the budget of the working group CAFF, with secretariat on Iceland (SAO meeting Yakutsk, 2005). However, a third explanation for reduced visibility – which most likely then later gave rise to a national strategy *on* visibility – reflects the years around 2007–2011 as the big player years.

When Denmark arranged the Ilulissat meeting in 2008, Iceland was not invited, and expressed concern: 'the issues to be discussed in Ilulissat were of general importance to all AC members,' Iceland argued, and requested everyone to be invited (SAO meeting Narvik, 2007). Despite Iceland's objection, the meeting was not only held but also followed by another meeting in 2010, this time with Canada as the organizer. Again, Iceland pointed toward a lack of inclusiveness and expressed disappointment over the fact that one more A5 meeting had taken place (SAO meeting Ilulissat, 2010). A third meeting in 2015, this time on unregulated fishing, once more told Iceland – through a lack of recognition – what it was not: an Arctic coastal state.

Iceland is often linked together with Sweden and Finland, due to the three of them being perceived as 'small' states, with lower expectations on leadership (Nord, 2016b). Neither do they attract any negative judgments or expectations from other members. In line with this, Iceland also performs with a certain degree of modesty. When Iceland ended its AC chairmanship in 2006, the state thanked all for support, and declared itself privileged in being able to build on the outstanding work of its predecessors: "[i]f we have succeeded, it is largely on account of the unselfish contribution of people too numerous to mention by name" (AC Ministerial statement by Iceland, 2006). At the next round of chairmanship, the message was close to synonymous; "you have made our job a lot easier," Iceland informed the predecessor Finland, and described the latter's chairmanship to have been 'outstanding,' there for Iceland to build on (ibid.; 2019).

### *What Social Interaction Has Brought About*

Iceland's role conception is ambivalent. On the one hand, Iceland conceives itself to be an Arctic coastal state, and the Arctic strategy also describes a state with knowledge and experience fitted the conditions of the region. It also identifies, for instance, shipping and resource utilization as opportunities – but also challenges – facing the Icelandic people. In these respects, Iceland's role conception is rather similar to the rest of the coastal states. On the other hand, this is a role conception that cannot operate freely since not even Iceland truly conceives itself to be a coastal state; 'Iceland has a small population, no oil industry and no technological know-how,' a state representative informed (interview 2015). Thus, when Iceland performs in the AC, it may engage in outreaching activities showcasing the Arctic Council, but it does not explicitly attempt showcasing *Iceland* as a coastal state, for instance, when assuming a big international project such as AC chairmanship (see Iceland's chairmanship program, 2019–2021). The lack of inclusion by other coastal states in the AC, most notably illustrated through A5 meetings, further has this perception reinforced. Thus, Iceland may await recognition as a coastal state, but in the AC it performs as a follower and steward, with ambitions.

## Norway

*National Arctic interest: to save the Arctic while making the most of its potentials.* The Arctic strategy of Norway contains an illustration over a variety of its Arctic activities: there is a fishing vessel and a fish farm, an LNG tanker and a rig, a tourist ferry and polar bears; a reindeer and some tents, to mention a few (2017).[11] The Arctic, or the High North as called by Norway, is, as explained by former Foreign Minister Børge Brende, part of the Norwegian identity: "[t]he Arctic waters are our natural home" (Brende, 2015). Norway is a shipping nation, with the fourth largest fleet in the world. In addition, it ranks as a global top five exporter of crude oil, where it also supplies the EU with about 20 percent of the Union's gas consumption (European Commission, 2017b). Although less than 3 percent of all recoverable oil and gas in 2011 came from the Arctic shelf, this figure is increasing: in 2016, the oil field Goliat went on stream, and additionally three more fields are preparing for production.[12] Apart from boosting Norway's northern economy, it is predicted that these fields will create about 3000 jobs (Oseland and Raspotnik, 2018).[13]

To Norway, it is described to be important to take environmental and climate considerations into account in everything that is done, and to be a leading nation as regards environmental policy (Arctic strategy, 2006:foreword, 45; Arctic policy, 2014). Its Arctic strategy ensures that growth and development does not increase the pressure on the environment, but rather takes place in such a way that greenhouse gases are reduced (2017:12). Indeed, climate ambitions in Norway are high, and by 2030 Norway intends to have achieved climate neutrality (Stortinget, 2016). In addition, Norway is ahead of other states when it comes to per capita share of non-emitting cars, and it also generates electricity from wind and water rather than fossil fuel, and many Norwegians spend their weekends close to nature, skiing or hiking.

To avoid a contradiction between these values of business opportunities (natural resources) on the one hand, and environmental protection on the other, Norway has combined them in a role conception of being a leader within green technology. Because, as Prime Minister Erna Solberg explains, a harvesting of Arctic resources requires green and innovative solutions (Solberg, 2017). Norway's efficient technology renders the country one of the most carbon efficient producers of oil and gas in the world, ranked also as a world leader on safety standards on offshore drilling (Equinor, 2020; Byers, 2013:205). Norwegian resource extraction is governed by a discourse similar to "drilling for the environment," where Norway considers itself to have the competence to extract in a responsible and environmentally safe way (Jensen, 2012). Thus, Norway's ambition would be no less than to be a leader in the field of knowledge (ibid.:27), which includes a responsible management of resources.

*The Arctic – linking the national and the international arena together.* Norway has yet a somewhat exposed position in the Arctic, as well as advanced Arctic interests. The country is a strong advocate of NATO, but it is also keen on having good relations with its neighbor Russia and navies regularly do exercise together (Arctic strategy, 2017; Bergh and Klimenko, 2016:64–65). However, previously, the Arctic policy has explained that relations with Russia can only be good to the degree they do not impact negatively on the Western relations (Offerdal, 2014:83). Despite 'resource muscles,' Norway is still a small state fostered to cherish values of cooperation, diplomacy, and mediation; its aim is to have the Arctic a stable, peaceful, and predictable region, where international cooperation is the norm (Arctic strategy, 2017:17).

When Norway is about to define its role in international settings, international voices are used to have the role conception supported: in Arctic policy documents, Norway is described knowledgeable and environmentally pro-active, as someone who shoulders cooperative leadership and raises awareness (Arctic policy, 2014:14).[14] To be a distinguished Arctic state is important to Norway, who also perceives it to be a responsible leader in taking the Arctic forward, relying on its long Arctic experience, its polar science, and its technological know-how. In the light of these conceptions, the Norwegian slogan found in its Arctic strategies – "The Arctic, Important for Norway, important for the world" – implies that the Arctic is linking Norway and the international sphere together. Relying on its Arctic assets, it frames itself as a nation helping out and wanting to do good, for instance, each day it 'feeds a lot of people' by producing well over 30 million meals of fish (Brende, 2017), Arctic gas is an environmentally friendly alternative to coal (Equinor, 2020), and the Barents Sea deposit can improve energy poverty and help the world meet its growing energy demand (Arctic policy, 2014). The selling point of the argument regarding fossil fuels would thus become a presented argument of environmental benefit *for all*. As such, the Norwegian role conception also manages to overcome the difficulty of explaining how offshore oil and gas extraction still could represent a responsible move forward toward greater environmental protection.

### Norway's Role Performance in the Arctic Council: Restless Know-How Leader

In the Arctic Council, Norway signals activity: 'the AC should deliver,' and Norway would be the helping hand. If one truly knows of something, like Norway conceives itself to do, then it is not that farfetched to also believe one could positively contribute to cooperation. Experience and knowledge obligate, and Norway would thus constitute a 'restless know-how leader' within the AC context that wants to see things move forward.

*Salient issues.* Early in the AC cooperation, Norway realized the presence – and effect – of climate change in the Arctic, pushing for it being the main theme of AC work (AC Ministerial Meeting Barrow, 2000; SAO meeting Oulu, 2002). "Global warming is a fact," Minister of Foreign Affairs Jonas Gahr Støre said when acting as the AC chair, proclaiming its man-made dimensions as beyond doubt (AC Ministerial statement by Norway, 2006a). Indeed, within climate change, Norway has performed vividly,[15] for instance, telling the other states "[t]here are things we can do and actually have to do – here in this region […]" (AC Ministerial statement by Norway, 2013). One such thing that Norway has informed the rest to do is the prohibition of routine gas flaring, believed to cause up to 40 percent of the black carbon in the Arctic (SAO meeting Fairbanks, 2016; Government of Norway, 2015a).

Equally early in the AC cooperation, Norway was quick to identify future development of oil and gas as a source for increased maritime transport, thus, also identifying a need for enhanced marine safety precautions (SAO meeting Rovaniemi, 2001). Oil and gas have been high on the agenda, at the same time as the improvement of safety clauses, implementation of oil and gas guidelines, and oil spill prevention are deemed vital (AC Ministerial statement by Norway, 2009; SAO meeting Copenhagen, 2011. More recently, Norway has revealed indignation over what is perceived to be a one-sided portrayal of oil and gas as simply involving risks (SAO meeting Fairbanks, 2016). That transport and maritime safety, environmental protection, oil and gas extraction as well as fishing can operate side by side in the Arctic is evidenced by Norway, the state has argued (ibid.).

Environment and energy are interrelated, as Norway sees it, and such an approach can be supported through integrated resource management: all future economic activity must take account of the fragile eco-systems, and the approach represents an opportunity to show how modern management and technology allow for commercial moving ahead while still protecting the environment (AC Ministerial statement, 2006a; 2006b). This is visible through the encouraging results from Snøhvit gas field, where the precautionary principle has been guiding extraction (SAO meeting Oulu, 2002; Fairbanks, 2016).

*Characteristics as an AC partner.* A guiding principle for Norway would be to 'grasp opportunities in a responsible manner' (AC Ministerial statement by Norway, 2013; 2015). "There is no such thing as a free lunch," the Norwegian minister let the new observers know upon their granted membership, simultaneously calling for a sound balance between commercial interests on the one hand, and social and environmental on the other (ibid., 2013).

As a state with plenty of Arctic knowledge, and plenty of Arctic interests, Norway can afford to act in a firm manner, take the floor to voice opinions, and, in general, take on leadership. At the same time, in other contexts, Norway is not a major (material) state but a small (normative) Scandinavian state, which also could help explain why Norway decided to stay silent on the matter of A5, and did not, in contrast to Russia and Canada, express its support thereof. It has also, similar to the Scandinavian countries, dedicated a fair share of its chairmanship to increase efficiency and institutional structure, for instance, by providing the AC with a permanent secretariat, based in Tromsø.

It has been a Norwegian trademark to avoid duplication and overlap in projects (SAO meeting Tromsø, 2007; Svolvær, 2008; Copenhagen, 2009b). Norway is a restless performer, who wants the AC projects to "[set] bold targets." It is calling out for concrete actions on black carbon reductions, possibly with even more ambitious common goals attached, and it wants political decisions to "move forward," relying on evidence, facts, and insights (AC Ministerial statement by Norway, 2006a; 2006b; SAO meeting Rovaniemi, 2018). This is a type of language that resonates well with the industrial parlance of best practices, harmonization, and exchange of experiences – a vocabulary appealing to reason and rationality rather than softer notions of feelings.[16] This would also characterize a restless know-how leader such as Norway, who, for instance, asks the U.S. to clarify why data on resources at risk for oil spill has not been delivered on time (SAO meeting Rovaniemi, 2001), or who quickly corrects the U.S.-voiced assumption that the AC is being an expert on oil and gas. "Wrong, the experts," Norway said, 'are found within petroleum producing states and their national authorities and research institutions' (SAO meeting Fairbanks, 2016). In those words, Norway described and legitimized itself as an Arctic state of weight.

### What Social Interaction Has Brought About

Norway's role conception is one of being an Arctic rescuer who knows how to manage Arctic activities in a responsible manner. Others' expectations are also supportive of Norway playing this role and why Norway faces no difficulties in performing a role consistent thereof: it performs as someone who knows about the Arctic and therefore is ready to act like a leader. When things are going too slow and cooperative progress is absent or not as complete as hoped for, the step is not far until a know-how leader gets restless. The things that Norway knows of and has an interest in are many, and its role conception is formulated quite loosely around a big Arctic role-set which leaves room for flexibility and adaptation to occur when interacting socially. In addition, the leader position upheld in the Arctic Council is not so much due to Norway naturally considering itself to be leader; rather, one acts as a leader since one has the knowledge to contribute. To Norway, to act as a leader comes perhaps less natural than it is practical, and its incentives seem therefore less to be about gaining absolute control over the cooperative direction and more to make sure that the very same direction is a balanced movement between business opportunities and environment protection.

## Russia

*National Arctic idea: the Arctic as a strategic resource base.* With the Barents Sea just within reach, the city of Murmansk is home to just over 300,000 people – less than during the Soviet days, but still vivid enough to signal a profound Arctic interest. The city is the home port for many fishing and shipping vessels, as well as some of those forty icebreakers that Russia so far has built and possesses. Also located here is a SAR center, the first of several along the Northern Sea Route. This route is envisioned as a future commercial opportunity of magnitude, should it turn into an international shipping lane for trade, with Russia employing its icebreaker capacity escorting foreign vessels through[17] (Granholm, 2016; Sergunin and Konyshev, 2016:82; Arctic strategy, 2008). Indeed, everything about the Arctic water is of a strategic importance to Russia; the surface for transportation/escort, the water column for fishing, and the shelves underneath for oil and gas (Arctic strategy, 2008).

Russia's Arctic policy differs somewhat from the rest of the states in its harder tone of national interests, sovereignty, and security. The Arctic is understood to offer development to the country (Arctic strategy 2008, paragraph 4a), where the region is sought to be turned into a strategic resource base. With an economy highly built around petroleum-related resources, Russia is the world's largest crude oil producer and second largest natural gas producer. About two-thirds of Russian oil and gas is already now Arctic, but it is estimated that as much as 95 percent of its remaining gas, and 60 percent of its oil, is deposited in the Arctic zone (Bergh and Klimenko, 2016:48–49). In December 2013, Russia became the first state to come on stream with an Arctic offshore oilfield: the Prirazlomnoye field located 60 km off the shore in the Pechora Sea. With a salient track record of accidents and spills, equivalent to more than two Deepwater Horizon leaks being spilled every year, increased oil production increases environmental risks further (Bachman, 2010; Luhn, 2016). Russia's main environmental priority would also be clean-up and preservation of unique ecological systems (Arctic strategy, 2008; 2013).

The indigenous communities have criticized Russia's resource extraction activities, arguing that environmental control has been lost to a third party, i.e., business interests.[18] Since then, a follow-up Arctic strategy has been developed (2013), dedicated to economic and social development. It has been suggested that the indigenous population might yet have a somewhat – relatively – stronger position than expected from an autocratic state (Hough, 2013:92–93). Because, although the Arctic above all is a Russian (material) asset, the Arctic is also highly symbolic: the Arctic strategy explains how mass media and culture could serve to nationally form a positive image of Russia, telling the story of Arctic development through Russian explorers (2008: paragraph 10d). In such a story, Russians would be people of the north; hyperboreans, 'beyond the North wind,' who are following their own ideals of morals and patriotism (Sergunin and Konyshev, 2016:35). The Arctic, then, becomes an area where Russia can broadcast to its audience a demonstration of national unity and international strength.

*Orchestrated spectacle on international strength.* Russia conceives itself to be major Arctic state, linking leadership to material power. Over the last couple of years, Russia has significantly increased its Arctic military strength and visibility, for instance, by establishing the Joint Strategic Command that command and coordinate all military units in the Arctic. Russia's Northern fleet has also got its own air force and defense, and military bases have both been reopened and built (Aliyev, 2019; Pettersen, 2016; Staalesen, 2015). In 2017, the Defense Ministry welcomed website visitors to take a virtual tour of the recently built military base – with several more to come – on the ice-covered island of Franz Josef, located far north in the Arctic Ocean. Here, 150 soldiers are supposed to be on duty, in temperatures as low as minus 40 degrees Celsius (BBC, 2017).

It might be that the larger international power-political picture has revealed 'irreconcilable differences' between Russia and the West (Carlsson and Winnerstig, 2016), but turning to Russian Arctic interests for guidance, we find that differences might not be that irreconcilable after all. Just like the other states, Russia emphasizes the importance of stability and strengthened relations in the Arctic (Arctic strategy, 2008: paragraph 6f; 7c). "We should maintain the Arctic as a region for peace and co-operation," President Putin has declared, pointing toward the interdependent nature of Arctic activities: "if you stand alone, you can't survive in the Arctic" (Harding, 2010). Moreover, should the Northern Sea Route be a lucrative business in the future, it takes shipowners willing to sail. The same is true especially for offshore oil and gas development, which suffers from an underdevelopment and lack of basic infrastructure, monitoring, and technology (Arctic strategy, 2013; paragraph 5b). Russia relies on foreign technology and expertise, why the economic sanctions imposed by the EU and others due to the Ukraine Crisis have had a negative effect on Russia's resource development (Astrasheuskaya, 2019; Rahbek-Clemmensen, 2017:5).

Russia acknowledges it cannot act alone and envisions a more regulated Arctic Council with legally binding agreements; the idea with a strengthened AC would be to increase control and reconfirm sovereignty as defined by the law of the sea (Zagorski, 2014:104). That the current military build-up should be used by Russia to expand territory is rather unlikely.[19] Instead, armament could be interpreted as an orchestrated spectacle on strength, possibly aimed at an audience outside the Arctic context. Indeed, although the Russian role conception is that of being a 'powerful actor' materially, it is well aware of not being there yet (Hønneland, 2016). Russia therefore needs not only cooperation but also symbolism in order to reinforce the image of being powerful: really, anyone who can break and master ice of three meter thickness, perhaps even five,[20] resembles someone as in control of the 'top of the world,' so Russia seems to argue.

### Russia's Role Performance in the Arctic Council: Responsive Informant

In the Arctic Council, national interests and respect for sovereignty are highly cherished by Russia, but rather than spelled out as tough demands in line with role conceptions, they are verbalized when Russia acts as an 'informant' – telling the others about Russian national activities and competencies. As such, Russia attempts to take control over the drama, and how it is *presented* therein. Over the last few years, Russia has toned down its informant characteristics, however, balancing its more assertive Arctic politics outside the AC. Russia, thus, is a *responsive* informant, adapting to the context.

*Environmental cooperation for the sake of national clean-up.* In line with its Arctic strategy Russia acts in support of oil and gas and maritime security, where its AC chairmanship, for instance, resulted in an undertaking of ensuring future sustainable oil and gas production in the Arctic (AC Ministerial statement by Russia, 2006). However, with the bulk of AC's work in terms of climate and environmental projects directed at Russia due to its high levels of contamination, Russia did early on express gratitude to WGs and states for understanding the importance of finding solutions for its environmental problems (SAO meeting Oulu, 2002). Within the area of climate change, Russia has even expressed a restless attitude on things going too slow: how would ACIA deal with thawing permafrost and its considerable problems caused to infrastructure (SAO meeting Inari, 2002); should not the AMAP also include the Russian proposal of the effect of climate change on the flow of Russian rivers within the current project (SAO meeting Svartsengi, 2003); and would not climate adaptation be the way forward for AC-related work (SAO meeting Stockholm, 2013)? However, to support scientific projects on climate change is not the same thing as supporting national emission reductions or other forms of comprehensive political actions on it. As such, although Russia has declared being on-board regarding reductions of black carbon, even planning for strengthened participation in AC work on the matter, and emphasizes its commitment to concrete projects like replacing fuel oil with LNG shipping (SAO meeting Whitehorse, 2015; Oulu, 2017; Rovaniemi, 2018), climate change is difficult for Russia, who prefers to talk about sustainable development.

*Characteristics as an AC partner.* In some ways, Russia is a leader, just as the Arctic strategies envision. This concerns, particularly, a project-type of leader within projects on maritime safety. However, in other aspects, like the leading of discussions, attracting followers, or employing organizational skills, Russia would not be a leader. Foremost, Russia is an informant who prepares everything to be said and done at home, and then informs the rest at the next meeting. 'Control' would be the key word, which also means to, under no circumstances, voice an opinion revealing some sort of discontent with any other actor. Instead, Russia maintains a 'high moral,' as should a hyperborean.

At times, Russia falls behind, not always performing in the same speed and with the same commitment as the rest. For instance, "[the Stockholm convention] will be ratified when everything is in place," Russia responded, when Sweden voiced concerns over Russia's possible lack of priorities (SAO meeting Yakutsk, 2005). Sometimes, however, Russia promises repercussions, like assuring that its state company Rostechnadzor's waste disposal in the heavy polluted Komi Republic will not occur again (SAO meeting Svolvær, 2008). At other times, cooperation is understood as going too far; when scholar Oran Young suggested the SAO's to enhance Arctic governance on sustainable development, Russia warned: "one should be careful to bring potential divisive ideas in the future AC" (SAO meeting Copenhagen, 2009b). Similarly, Russia has opposed phrases such as "strengthening governance of Arctic," since translated into Russian these will have an "unintentional meaning" (SAO meeting Ilulissat, 2010). Following its role conception, cooperation could thus never trespass the border where Russia will lose national control.

Compared to other states, positive steps taken by Russia are in higher degree noticed and paid attention to by others. For instance, the U.S. is appreciative of Russia's "expanded and vigorous participation in the AC" (SAO meeting Washington, 1999); Norway congratulates Russia in progressing on PCBs and noting it as an important member of the Council and Arctic state (SAO meeting Fairbanks, 2000; 2016); and Finland highlights Russia's work on marine protection as important to all of the Arctic (SAO meeting Reykjavik, 2003). States – most notably Norway and the U.S. – are boosting Russia's cooperative commitment by confirming and recognizing it as important.

### What Social Interaction Has Brought About

Russia's role conception is 'soon-to-be leader of top of the world.' In the Arctic Council, however, it performs less like a leader and more like a (sovereign) informant, who wants to be in control. Since cooperation is key to national interests and future leadership ambitions, a twist is added through being a 'responsive' informant that is rather adaptive to the context. Within the 'informant' role performance also lies embedded what has been referred to as Russian story-telling, where, possibly, the Arctic is told "more Russian than Russia itself" (Hønneland, 2016:176).[21] The Russian story includes things such as caring for science, while also being a major Arctic state. From this perspective, the symbolic value of actions should not be underestimated. For instance, the Russian role conception is confirmed when the state is hosting a major Arctic scientific event like the International Polar Year (SAO meeting Svartsengi, 2003; Yakutsk, 2005), or, when standing on the North Pole, a photography is taken of all of the Arctic SAOs, the flag of the Arctic Council, a Russian state official, and explorer Chilingarov (AC Ministerial statement by Russia, 2013).

To Russia, cooperation is needed if its Arctic interests should come true. This requires a toning down on unilateral behavior, also when performed by others. Illustratively, a Canadian crewless barge had drifted from the coast of Canada, through the U.S. and into Russian waters where it was stranded for a while. Contextually, it was in the middle of the Ukraine crisis and all the political cold it brought about. Russia described the episode of having it returned to Canada as 'good cooperation' (SAO meeting Anchorage, 2015), despite taking nine months of drifting until it was back in Canada, with little or no efforts by Canada to track and tow it. To tell stories and present (positive) images on good cooperation – true or false – is, however, a way for Russia to strike the correct balance between cooperation and national interests, by convincing the rest that they are performing the role of a reliable cooperative partner. To succeed, Russia must perform a responsive role that pays attention to, and navigates between, social norms and others' expectations.

## Sweden

*National Arctic interest: to spread scientific knowledge.* With just some 100 major icebreakers in the world, Sweden with its five state-owned vessels is ranked globally top three regarding icebreaking capacity, just like Finland.[22] Amongst Sweden's icebreakers is *Oden*, a powerful icebreaker designated to polar waters. It is foremost a research platform firstly available to the Swedish Polar Research Secretariat, but often in cooperation with other countries, for instance, the U.S. who considers it an important polar research component (Majlard, 2016). Previously, *Oden* had been chartered for oil and gas prospecting through mapping of Arctic sediments, but for some time no leasing was allowed for purposes involving oil prospecting (Swedish Maritime Administration, 2012).

For Sweden, the Arctic is about research. The Swedish government describes the country as conducting Arctic research of highest rank, even to be world leading in polar research (Arctic strategy, 2011). When *Oden* soon is to be retired, a future without Swedish icebreaking research is therefore assessed to not only cause Sweden's polar research "irreparable damage" but also its "reputation within the international research area [author's translation]" (Swedish Polar Research Secretariat, 2016). The planning has started for a new and even more powerful research-dedicated icebreaker, with an improved environmental performance like, possibly, being powered by methanol as the first of its kind (Carlsson, 2019).

To Sweden, three Arctic key priority areas are climate and the environment, economic development (foremost, forestry and mining), and the human dimension. Concerning the latter, Sweden – just like Finland and Norway – is committed to the preservation and development of possibilities for the Sami people to maintain their traditional livelihood (Arctic strategy, 2011:42). The negative impact of climate change on reindeer husbandry is extensive, and augmented pressure on human health is furthermore a risk effect from high levels of organic toxics, like PCB and heavy metals (mercury). Russian industries in the Arctic are pointed out as the main pollutants (ibid.: 44).

*Science – connecting the national to the international.* The environmental know-how of Sweden would be its main selling point in the Arctic operations, but it also identifies a competitive edge within the maritime industry. The environmental field is, however, an area where it states itself to be world leading, and intends to keep such a position (Arctic strategy, 2011). Here, it has a proper chance to contribute internationally, and to strengthen its reputation as an advanced green state. As a state, Sweden consistently ranks high, even highest, on the CCPI (2020).

The strategy lists four areas that Sweden should work for in the Arctic Council: substantially reduced emissions of greenhouse gases and climate pollutants, a strengthening of Arctic resilience and adaptation, increased use of environmental assessments and environmental mappings, and reductions of transboundary pollutants. In a supplementing Arctic policy from 2016, which addresses the Arctic environment only, Sweden confirms these areas while ensuring to work to strengthen the environmental dimension of the Arctic Council (2016). Sweden's role conception is one of being a knowledgeable environmental actor who can contribute internationally by applying an issue-specific perspective in a neutral manner.

### Sweden's Role Performance in the Arctic Council: Teacher on Demand

The role performance of Sweden in the AC is one of an environmental teacher. It is a role that is performed only if there are pupils willing to learn, which indicates the role to be a teacher on demand.

'Demand for' is an important dimension of Swedish role performance. For instance, before assuming the chairmanship, but also after, Sweden was very much in the background when member states were gathered, causing several member states to openly wonder what Sweden's stance on different issues actually might be, since it provided for so few opinions on AC priorities (Nord, 2016b:57). *During* the chairmanship, though, the state took a giant step forward in terms of leadership, visibility, and determination. Sweden shouldered a role and script familiar since before and performed appropriately in line with this position. Sweden has a long and successful record in global diplomacy, and in the AC, it also played a significant role in terms of pushing AC cooperation forward (ibid.: 58–59).

*Salient issues.* Sweden follows the same pattern as the other two non-coastal states, and in the SAO meetings, reveals lower frequency rates on issues spoken about: it is (almost) all about the Arctic environment, and especially things such as short-lived climate pollutants, biodiversity, and Arctic resilience. On climate change, Sweden has called out for "an even stronger environmental dimension in the Council," and referred quite early on to short-lived climate pollutants as an area "where we, the Arctic countries, can take measures that will directly benefit the Arctic region" (AC Ministerial statement by Sweden, 2011; 2015). Within the AC context, climate change and oil extraction are considered as a dilemma, being causally linked to one another. To label it as a dilemma is telling of a pragmatic and mediating perspective on how to cope with economic reality while still moving forward environmentally.

*Characteristics as an AC partner.* As explained by Gustaf Lind, former SAO and chair, the task for the chairmanship is to search for commonness and things that unite. The logic behind international cooperation then becomes to respect those different positions about which an agreement cannot be reached on – or at least seems impossible to agree on – and instead focus on pushing the common agenda forward (Lind, 2011). Here, Sweden succeeded quite well in terms of continuing the work started by Norway and Denmark in strengthening the Council's institutional structure and in moving specific issue areas forward, e.g., resilience. The lack of Arctic Ocean territory made the other states view Sweden as lacking a bias, since no strong interests were expected to color its leadership; it was 'harmless' (Bergh and Klimenko, 2016:66). However, to act and be perceived as an honest broker also legitimized Sweden's attempt to create an enhanced understanding of Arctic issues.

Sweden is not an AC member that speaks more than necessary, unless the issue at hand is familiar and one where Sweden can contribute. For instance, it is active in resilience; leads issue-specific task forces on short-lived climate forces as well as scientific cooperation; and advocates for making women heard in scientific as well as political processes (cf. SAO meeting Espoo, 2001; AC Ministerial meeting 2000; AC Ministerial statement 2015). These are all signatory issues where Sweden represents the stance of taking environmental cooperation a normative step forward.

"The Swedish example shows that it is possible," Foreign Minister Margot Wallström stated, referring to how emission reductions and increased GDP indeed is possible (AC Ministerial statement by Sweden, 2017).

But to act where one senses a call for expert knowledge also pertains to issue areas outside the environmental field. As an example, in the first available transcript of a SAO meeting, Sweden kept a low profile. When the remark eventually came, it was information on an upcoming icebreaker project on "Tundra northwest" (SAO meeting Anchorage, 1999). Sweden has then continued to show expert-knowledge on this, where a leaning on its knowledge became a safe haven when uncertain on its role and contribution. However, to offer help and expertise needs to be balanced, like when Sweden was quick to follow up its offer on icebreaker expertise in relation to Arctic SAR, by pointing out that "we can learn a lot from each other" (AC Ministerial statement by Sweden, 2009).

What Sweden offers the AC is mainly knowledge, but within its role performance lays also a reluctance to act prior to a request thereof; a certain degree of common understanding should be sensed. Since common understanding grows out from communication, communication is central for any outreach activity as 'facts do not speak for themselves' (AC Ministerial statement by Sweden, 2011; SAO meeting Copenhagen, 2011). It is no coincidence that Sweden, at a time when the Internet was still new, was hesitant to the launch of a new AC webpage as an information hub: if people do not get a paper in their hand, they will not read it, Sweden warned (SAO meeting Fairbanks, 2000). Efficiency, transparency, and clarity have been highly ranked, for instance, by developing reporting templates for WGs; to launch 'Two-pagers' to allow the public to be aware of the AC discussions; and to urge for meeting agendas to be 'short and focused' (SAO meeting Luleå, 2011; Haparanda, 2012). Furthermore, to make sure that confusions are avoided and that everyone *understands* – for instance, *why* resilience is important (SAO meeting Stockholm, 2013) – describes Sweden's role performance as a teacher wanting its class to progress.

### What Social Interaction Has Brought About

Sweden's role conception is that of someone who cares for the environment, believes in science, and knows the topic spoken. Ultimately, it is that of an expert who could teach others. Yet, its role performance reveals that teacher to be highly attentive toward others' wishes for 'education.' Illustratively, when the U.S. presented a discussion paper on climate change, noting "ACs potential role in educating the world about the Arctic and climate," Sweden would be the first to offer its support (SAO meeting Fairbanks, 2016; SAO discussion paper, 2016b). To act as a teacher on demand implies being sensitive to when it is appropriate to teach, and when others are susceptible to be taught. In the case of Sweden, social interaction thereby contains important information regarding when there would be an international need for a knowledgeable teacher to push environmental governance forward.

Sweden works with rules, implicit as well as explicit, instead of pushing its agenda through. Critique against Russia as a major source of pollution in Sweden was, for instance, not shared until Russia chaired the AC, i.e., was 'in power' and more legitimate to criticize (SAO meeting Yakutsk, 2005; Khanty-Mansiysk, 2005). However, after all, it could be that the critique included more of a profound personal dimension. Because, in a timely context, Russia's strict control over its Arctic passages had made it difficult for Sweden to go through with its research expedition: samples could not be taken out from Russia, and *Oden* was required an (expensive and time-consuming) escort by the Russian icebreakers (ibid.). If Sweden could not pursue research from a research platform like *Oden*, Sweden would be deprived of what makes the country unique in international settings. It would simply risk losing the confirmation from others of being a knowledgeable actor, and as such, put the Swedish role conception in jeopardy.

## U.S.

*National interest: to be a major maritime nation (but not necessarily 'Arctic').* The U.S. gained status as an Arctic state through its purchase of Alaska from Russia in 1867.[23] Such a – in the context – recent Arctic acquisition also renders the American feeling of being Arctic rather weak, disconnecting the 'Arctic reality' of those living in Alaska (Restino, 2015). Illustrative, when former president Obama travelled to Alaska in 2015, he became the first president in office to officially visit Alaska (The White House, 2015).

The U.S. is a maritime nation. More than 5,000 large vessels and cargo ships each year pass the Unimak Pass at Alaska's Aleutian Islands, including heavy oil transportation from Canada. Also, the Bering Sea is home to U.S.' biggest commercial fishing asset, valued two billion U.S. dollars annually (Byers, 2013:159–160). In line with this, the Arctic policies and strategies of U.S. describe security, safety, and stewardship at sea to be its focus area. Arctic activities should furthermore proceed on the basis of the best available information and respond effectively to environmental challenges (Arctic policy, 2009:9; Arctic strategy, 2013:7–8). However, over time, a discrepancy in the administration's interpretations of such environmental challenges, not least climate change, is noticed: while President Clinton signed the Kyoto protocol, the U.S. Senate refused to ratify it; whereas President Obama advocated a strong Paris agreement, President Trump withdrew it; and whereas Obama enacted a law to invoke an indefinite ban on U.S. Arctic offshore oil and gas drilling (Fears and Eilperien, 2016), President Trump signed an executive order declaring the U.S part of the Arctic Ocean reopened (Reuters News Agency, 2017).

The people living outside the urban areas of Alaska have for long experienced environmental pollution, while also searching for infrastructure, development, and economic stability. About one-third of all jobs in Alaska is connected to the O&G industry, where the onshore oil field, Prudhoe Bay, discovered in the late 1970s, would be one of the most productive oil fields in the Arctic in total.[24] Oil tax revenues fund most of the Alaskan government services (Alaska Oil and Gas Association, 2017). To scale down fossil oil extraction in a region estimated to hold not only the highest quantity of deposits but, according to the USGS, also highest amount of easy extraction therefore is not necessarily supported by the people.

*Science – an international unifier.* National security is a top Arctic priority; the U.S. conceives itself to be great power, who cherishes its sovereign rights and territory. Freedom of navigation, under, over, and through air spare and Arctic waters is vital to its national security, in addition to safeguarding national interests (Arctic strategy, 2013:2).[25] However, so far, allocated resources have not led to any major strategic investments in the Arctic: there are no military installations north of the Arctic Circle, nor any SAR centers bordering the Arctic or the Chukchi sea (Hilde, 2014:156–157; Alaska Rescue Coordination Center, 2017). Furthermore, the coast guard has for years relied on two icebreakers (of which one is broken), while requesting for six new vessels. Only recently has congressional funding been secured, enough to sign a contract for the first U.S. heavy icebreaker in over forty years (Humpert, 2019).

Security interests are also linked to softer issues of commercial and scientific (environmental) character, which makes AC cooperation attractive and important, not least to advance safety and security in the region (Arctic strategy, 2013:9). In 2009, a leaked cable from the U.S. Embassy in Russia to the State Department was reporting that trust and confidence would be two positive side effects of increased scientific cooperation (Byers, 2013:158). In that sense, environmental protection brings good values to Arctic relations. The U.S. conceives itself to play a leadership role in Arctic research, to be an "Arctic science power" (Conley and Rohloff, 2015:xiv) that dedicates resources toward greater understanding of Arctic conditions and why it also ascribes some of the positive AC results to its own scientific agencies (Arctic policy, 2009). Even though the current Trump administration has dismissed Arctic human-induced climate change, the U.S. budget pertaining to Arctic research and science has remained close to intact (Conley and Melino, 2019). For a state with weak Arctic identity, and whose Arctic policies are more descriptive than concrete on actions (ibid.; David, 2013; Offerdal, 2014), the conception of being a 'science power' gives role coherence.

### Role Performance in the Arctic Council: Innate Leader

The U.S. is a visible actor in the AC, not afraid of voicing opinions and aware of how to organize meeting procedures. As a project initiator and visionary, it has been less active and assertive. This reflects an actor that for sure performs as a leader, but less so because of having clear ambitions within Arctic cooperation generally and Arctic issues specifically. Instead, the U.S. role performance seems to find its nourishment from experiences outside of the AC where it is acting as a global leader, which in the AC then gets transformed into acting like an 'innate leader.'

*Issue-specific performance: late awakening but possibly increasing?* Initially, the U.S. was a reluctant AC member who constrained cooperation and caused some nervousness on behalf of others regarding their level of commitment (Finkler and Kadas, 2016; Nord, 2016b:45). However, during its first chairmanship 'it yet delivered,' and has since then continued to develop both an Arctic awareness and a positive view on the AC as well as the role it plays therein (Finkler and Kadas, 2016).

Maritime issues in the AC are important to the U.S., just as the Arctic strategies prescribe. But already from the onset it has also been active and involved in issues concerning climate change, described as an issue the rest of the world should be aware of (SAO meeting Washington, 2000; AC Ministerial meeting Barrow, 2000). However, 'behind the scenes,' U.S. behavior has been, at least in the beginning, less convincing, where the state was considered to put the ACIA process in jeopardy (Nilsson, 2007).[26] The U.S. mentioning of climate change also dropped, correlating with a conservative administration that pulled back from the Kyoto protocol. It dropped, but it was never completely abandoned and instead the U.S. became – together with Russia – the first country to present initiatives on black carbon within the Arctic Council (SAO meeting Svolvær, 2008; Kautokeino, 2008). The Obama administration stepped up climate stakes further, to the point Secretary of State John Kerry declared science on climate change to speak for itself; "It does not take a PhD to know that the combination of heat and ice produces melting," Kerry said, and recommended the meeting to take a look in the mirror to know where the problems began (AC Ministerial statement by the U.S., 2015). In the SAO setting, the state has explained its domestic climate policy as now being under review, premiering affordable energy and economic growth (SAO meeting Rovaniemi, 2018). The cooperative behavior of the U.S. seems to set the course back to the beginning of AC cooperation and unsure levels of climate commitment.

*Characteristics as an AC partner.* In the AC, the U.S. is a pragmatic and realistic leader, who wants to see order within cooperation and who prefers 'facts on the table.' Rather than having visions, it concentrates on what is possible to achieve, often with reference to itself as an actor. For instance, Finland was thanked for putting together a document in relation to the World Summit on Sustainable Development in Johannesburg, but it was unlikely that the U.S. would agree to have its name on it (Rovaniemi, 2001). Similarly, the U.S. stalled the otherwise consensual support on Norway's climate project, due to Norway's late distribution and what was perceived to be obscure project formulations (SAO meeting 2007, Tromsø). The U.S. is not afraid of bringing forward constructive criticism or to *instruct* on how things should be carried out, for instance; WGs have failed to include corrections in the reports, information delivered has been unclear, and issues have been addressed beyond their mandate (SAO meeting Inari, 2002; Tromsø, 2007). Science, in the U.S. opinion, is political dynamite to be handled with care. With that said, its role performance is still that of someone who aims for cooperative progress, who has the ambition to take learned AC lessons into account, and who leads by example – hoping its own behavior can inspire others to follow (see, for instance, SAO meeting Anchorage, 2015).

To act like an innate leader does not mean to not care for its reputation. Thus, when Canada in the early days of AC cooperation exerted considerable pressure on the U.S. to take on the chairmanship, i.e., to show good faith in cooperation (Nord, 2016a:25–26), the U.S. proved itself tangible toward such expectations and took on leadership. Still, the commitment was not perceived as full-hearted, and Norway restlessly commented on the long period of time that passed between meetings. Characteristically, the U.S. replied: "United States would evaluate its success as Chair not by the frequency of meetings but by the progress on the Council's program of work" (SAO meeting Anchorage, 1999). The U.S. role performance reveals a leader that does not like to be told what to do. Indeed, an innate leader can afford to be a righteous leader, i.e., someone who encourages others rather than offers criticism, and someone who 'listens, and listens more,' to get everyone onboard (SAO meeting Whitehorse, 2015; Anchorage, 2015; AC Ministerial statements by the U.S.). But an innate leader can also appreciate the view of others and listen to their concerns, like on issues such as climate change, without being rushed to make decisions. Rather, to the U.S., focus would not be anything less than "to make the right decision," as told by the Secretary of State (AC Ministerial statement by the U.S., 2019).

### What Social Interaction Has Brought About

The U.S. role conception is telling of a major maritime nation, who makes an important scientific contribution to Arctic cooperation. It is, however, not necessarily sure of what it is that the Arctic wants and what to get out of Arctic cooperation. In a context filled with states being 'Arctic experts' or various kinds, with knowledge and clearly articulated interests, U.S. leadership appeals more to the act of cooperating: the U.S. is keen on having clear communication, order within the administration, and harmonized and feasible projects, while not being afraid of voicing firm opinions and – depending on the context – to both push the breaks or the gas, for instance, in climate-related matters. Thus, the role conception of the U.S. may be that of a great Arctic science power, but in terms of role performance it does resemble more of an innate leader who leads more out of a habit than of issue-specific conviction. To act like an innate leader is also what other states seem to expect out of the U.S., further reinforcing a role performance in line thereof. However, the U.S. seems to have sharpened its interests and come to increasingly step forward on issues such as marine safety and black carbon, although, not without setbacks.

## The Stability of Roles in the Arctic Council Cooperation

This chapter has described the Arctic states' roles in the AC – their Arctic interests and their conceptions of them being a specific type of Arctic actor, compared to their performative pattern in the AC *(see Table 4.1)*. It has provided for reasoning around social interaction, and how this adds layer(s) to state behavior and wants. As one of its conclusions, this chapter now suggests that engagement in real interaction, and not only conceptions thereof, influences state behavior. However, an additional conclusion would be that a state's role and role behavior is rather tied to the script, following the trajectory of past and present role expectations and leaving little room for spontaneous deviations going in the opposite direction.

Although there are differences, this chapter did not reveal any major discrepancies between role conceptions and role performance (see Table 4.1). This supports the role theoretical argument of a role as something that is less based upon impulsive 'here and now' and more rooted in thorough reflections of proper behavior, given both the social setting *and* domestic expectations. Therefore, the differences that have been revealed between role conceptions and role performances concern a toning down of national rhetoric and interests in favor of Arctic cooperation and good relations. Social interaction polishes off some of the unilateral interests and solutions found within role conceptions and pushes role performances toward enhanced commonality. Thus, an effect of social interaction on roles would be the provision of, in comparison to role conceptions, a more low-key role performance. From this conclusion the distance is not far to another role theoretical key assumption, namely, the conclusion that *others matter*. The fact that role performance is partly a product of both specific and generalized others' and their notions of what cooperatively would seem to go, suggest that international cooperation – through the operating mechanism of socialization – has the potential to, in the longer run, bring somewhat enduring change to international relations.

Perhaps the most striking illustration of cooperation as gaining the upper hand in the role-taking process would be Russia's role performance during the political phase of the AC development (identified in Chapter 2 as the years 2006–2013) in comparison to its more withdrawn performance during current phase of cooperation under insecurity (years 2014–). Since much of this insecurity is sprung from Russia's behavior *outside* of the Arctic, Russia's behavior *within* the AC could be read as an attempt to tone down its unilateral appearance outside the Arctic and play by the rules of being a cooperator. In a context where calls for enhanced Arctic governance echo side by side with calls for enhanced commerciality, the Arctic states share a common understanding of the Arctic future as to hinge on

*Table 4.1* Role Conceptions versus Role Performance – The Effect of Interaction

| State | Role conception | Role performance | The effect of social interaction |
|---|---|---|---|
| Canada | (Tough) Protector of the North | *Protector* | Social interaction softens the protector role and pushes role performance in the direction of multilateralism |
| Denmark | Global Arctic player | *Pragmatic (coastal state) kid-brother* | Social interaction translates global Arctic player conceptions into a performance on pragmatism and responsibility, displaying a major player in the 'small' |
| Finland | Arctic expert | *Reserved team player* | Social interaction informs the state about an occurring need for expertise, and activates the expert role through cooperative cues |
| Iceland | Arctic coastal state | *Follower with ambitions* | Social interaction reveals the support of the coastal state role inadequate, thereby restricting role performance although ambitions remain |
| Norway | A knowledgeable Arctic rescuer | *Restless know-how leader* | Social interaction reveals a symmetry between role conceptions and others' expectations, which enables a strong, even restless, role performance |
| Russia | Soon-to-be Arctic/global leader | *Responsive informant* | Social interaction illuminates a need to 'play by the rules' and to play down unilateralism and world leader rhetoric, should advantage of Arctic cooperation be drawn |
| Sweden | Teacher | *Teacher on demand* | Social interaction displays the teacher role highly attentive to cooperative cues, like others' wish for being taught |
| U.S. | Major maritime nation | *Innate leader* | Social interaction provides for a performative direction, clearer than role conceptions would have suggested, and offers a role script familiar from other international settings |
| **Conclusion** | | | *The interaction effect targets actors' cooperativeness by displaying 'others' to matter* |

the way cooperation evolves. This goes for environmental and commercial issues alike. It is suggested that interacting states, not just Russia but all of the Arctic states, over a long period of time have been socialized to the notion of 'cooperation' as representing the appropriate way of organizing Arctic relations; that is, the Arctic states have all learnt how to perform the role of an Arctic (cooperating) state. By the same token, their performance contributes along these lines, reinforcing such an interactive structure.

Another type of illustration where role performance seems to grow out of interaction would be the U.S.' understanding of what it means to be a leader within the AC. As the state – through interaction and increased levels of trust among participants – became more aware of its Arctic interests, it employed leadership beyond the initial focus of displaying organizational skills onto an issue-specific leadership. As national role contestations through current political administration again view (climate) science as rather incomplete, a return to the leadership role once again constructed around organizational qualities seems likely. Similarly, as a result of interaction, once Sweden was sure of which rules that were 'governing the game of' Arctic cooperation, it became a more visible actor trying to push things through: consequently, by then Sweden was aware of when there would be a demand for 'teaching' or not, and the chairmanship offered clear suggestions of a demand thereof. Indeed, throughout all chairmanship periods, the state holding the chair has revealed 'more of its role,' i.e., has geared up on visibility and the level of action-taking, thereby responding to prescriptions of what to expect from a chairmanship.

The descriptive analysis of the Arctic states' roles in the AC has revealed roles to be a flexible construction. As new thematic issues have been added, roles have adopted to new norms; for instance, no state opposes either oil and gas activities or climate cooperation, despite a role conception favoring one over another. This inbuilt flexibility, issue-wise but also behavioral, is also what keeps the role stable by allowing it to adapt to contextual changes – new cues and demands – without having to abandon its role conceptions. Sweden acts like a teacher *on demand*, Russia like a *responsive* informant, Iceland as a *follower* with ambition, and so on. Roles are simply designed to allow for flexibility, where the most suitable part of the role-set is identified with help from social reflections on what reactions – real or perceived – they get from other AC members. As a result, although 'the thinking through others' are partly contributing to also role conceptions, experiencing these expectations 'in real life' tends to sharpen their behavioral impact. Social interaction, through the role performance dimension, is ultimately adding more flesh to processes of state learning. In the next two chapters, analytical focus will be concentrated on such reflective processes carried out by negotiating state representatives, illustrating how social interaction helps states to choose performance from their role-set.

However, lastly, in this chapter, the impact of social interaction on roles has also been discussed in the flexible way that states perform their – already – flexible role. Sometimes, for instance, Denmark would be more 'pragmatic' than a coastal state kid-brother, Finland more 'reserved' than a team player, and Russia more 'responsive' than an informant. And all is as it should be, given the many layers of social attentiveness that is lying imbedded within roles, making them more refined than always following the same script.

## Notes

1 The role mapping is conducted as a text analysis, relying on a code book to categorize and count different state characteristics according to three areas: (1) attention devoted to specific environmental issues, foremost climate change and oil and gas issues; (2) expectations and judgments on other fellow actors; and (3) behavioral characteristics, such as leader, initiator, expert, administrator, and so forth. Since the note-taking style of the AC has significantly varied over the years, where the first handful of years were very detailed but then became less detailed and more anonymized, the analysis is carried out in a relational manner, interpreting each state behavior in light of the frequency level revealed by the other states in the same period of time. Therefore, rather than attempting to establish behavior in numbers, the text analysis is dedicated to tendencies and patterns, which is illustrated not by numbers but through characteristic and personal examples of how the role is performed. Each meeting has also been qualitatively captured in written and divided per state, to provide the pattern with further empirical substance.

2 Meeting minutes are available from the year 1999. Minutes until the year 2016 to the number 33 was originally included in the scope of this study. However, upon the revision and update of this book, meetings up until 2019 have been included in the qualitative text analysis – not to quantitatively compose the role pattern but to have it further illustrated. Thus, for the role-mapping analysis, a total of 39 minutes is considered.

3 See also Chater (2016), for an overview of project sponsorship, accessed through the SAO reports, where Canada, Norway, Russia, and the U.S. are the most frequent sponsors.

4 Canada and Russia are both claiming certain segments of trans-Arctic shipping routes to be their internal waters. As internal waters, it means that national jurisdiction will apply, which would maintain those stricter rules on vessel construction, equipment, and manning that both Canada and Russia have adopted, compared to those established in the IMO instruments (Stokke, 2014:131–132).

5 Compare, for instance, Climate Change Performance Index (CCPI) of 2010, 2015, and 2020, where Canada ranked 59th, 58th, and 55th, respectively, of a total of maximum 61 countries investigated.

6 Canada is involved in more legal disputes, although none of them are causing threats to security – one dispute is with Denmark over the island Hans Island; another is with the U.S. on where territorial borders should be drawn in the Beaufort Sea; and a third one concerns the Lomonosov Ridge and the North Pole (Byers, 2009; 2013).

7 With the population of a little less than 57,000 inhabitants, Greenland is continuously dependent on Denmark for financial backing-up as well as on things such as personnel too, for instance, patrolling internal waters. Political areas such as defense and security policy, foreign affairs, and monetary systems are for now kept under Danish jurisdiction.

8 Denmark has been mapping and collected data about Arctic continental shelves since 2002, albeit with increased efforts starting in 2006 (Arctic strategy, 2011:15). The dispute on Hans Island constitutes the only dispute over land in the Arctic, and is an island consisting of 1.3 km$^2$ rock located in the Kennedy Channel between Canada and Greenland.

9 Actually, Finland showed the way (as it did with AEPS) by introducing gender equality to Arctic cooperation, eagerly cheered on by Sweden, Iceland, and Norway (SAO meeting Rovaniemi, 2001).

10 The Spitsbergen Treaty from 1920 establishes the – until then – terra nullius archipelago of Svalbard to fall under Norwegian sovereignty. However, the treaty made sure that the territory could not be used for military purposes and naval bases, though providing other states the right to continue with economic activities on Svalbard, as well as in its territorial water. Diplomatic friction has arisen since Norway argues that the rights of other states to resources such as fish, minerals, and possible oil and gas does not extend to the whole of EEZ, or continental shelves. Iceland has therefore not been accorded fishing quotas, despite being part of the Spitsbergen Treaty (Byers, 2013:19).

11 Norway has released three Arctic strategies and policies; in 2006, 2014, and 2017. The current strategy is titled *Between Geopolitics and Social Development* and keeps intact overarching Arctic goals from previous strategies and policies but elaborates more on sustainable development and how to achieve a regional balance throughout the country, focusing on its northern parts (2017:15).

12 Altha/Gota, Wisting, and Johan Castberg, where the last is expected to come into production in 2022.

13 One of these three fields, the major offshore field Johan Castberg, became accessible as a result of Norway recalculating the latitude of its average ice-edge, which was decided to be drawn further up due to Arctic warming and ice-melting. Norwegian safety precaution measures assure that offshore petroleum activity occurs with no less than 50 km distance to the imagined ice-edge, and Johan Castberg makes it just within the new ice-edge (Government of Norway, 2015b).

14 For instance, U.S. Secretary of State John Kerry, and Rajendra Pachauri, chair of IPCC, say states and others should be united in a concern for the region, and where technological capabilities are sought after.

15 Symbolically (i.e., theatrically), at the AC Ministerial meeting in Tromsø 2009, Norway invited former Vice President Al Gore – well-known for his demands for action on climate change – to give a presentation at the meeting's opening (SAO meeting Copenhagen, 2009a). Coalescing in time, the world was awaiting COP 15 and important discussions on post-Kyoto. The Norwegian chairman-ship therethrough saw a chance to make the Arctic region even more visible in the global climate debate.

16 This is a conclusion based upon the participation as an observer in industry conferences, where the science relied upon by the industry was repeatedly cherished for its objectivity and reliance on "hard" facts (The Arctic Oil and Gas Conference, 2014; Arctic Technology Conference, 2015; and IEA Gas and Oil Technology, 2017).

17 Full transit passages have so far been modest: in 2012, 46 vessels transited, to significantly decrease in the coming years, to levels below 20 transit passages a year. The 27 transit passages made in 2018 thereby represent a figure going up (Centre of High North Logistics, 2020).

18 It has been speculated that the decision of the Ministry of Justice to suspend RAIPON from participating in the Arctic Council for six months had less to do with the official explanation of irregularities found in RAIPON's organizational statutes, and more to do with RAIPON's critique of resource extraction in the North (Conley and Rohloff, 2015:99; Staalesen and Nilsen, 2012). RAIPON was permitted to reopen in March the following year, but there was suspicion of Moscow having interfered regarding the election of the new organization president (Conley and Rohloff, 2015). Recently, similar critique has again been raised, where Russia is claimed to remove critical voice concerning Arctic development (Nilsen, 2019).

19 The Arctic strategy refers to international law as the guiding principle of Arctic relations (2008: paragraph 7a), and this principle has previously been adhered to, for instance, in 1990 the U.S. and Russia settled a maritime boundary in the Bering Sea; and in 2010, Russia reached a similar agreement with Norway in the Barents Sea, which allows for fishermen to fish on both sides of the border as long as the total fishing quota is not exceeded.

20 Nuclear powered icebreakers – stronger than diesel-powered vessels – are unique to Russia, who has commissioned the building of a fifth one: the 'monster icebreaker' *Lider,* capable of breaking through five meters of thick ice it will escort ships and LNG carriers through the Northern Sea Route (Staalesen, 2016).

21 This phrase 'story-telling' is used by Hønneland (2016) when describing how all actors change and modify real events into stories. Since the story told – how it is received upon by those listening – is the one that provides actors with legitimacy, and not the actions or events per se, actors may sometimes try to adapt their actions to fit the (legitimate) story. In this book, whenever an actor is thinking of how to be perceived, the possibility of story-telling would arise.

22 In 2016, Finland added to its ice-breaking fleet the first LNG-powered icebreaker in the world, also considered the most environmentally friendly to operate in icy waters. Finnish vessels have been operating in Arctic waters and used for commercial purposes, where oil and gas companies have had them chartered. In a protest against Arctic oil extraction, environmental activists have boarded the ship Fennica, and in 2015 the so called 'kayaktivisits' protested against the ship being headed for Alaska under the Shell logo, thus, preventing it to head north (Brait, 2015).

23 Russia was reported to fear a war breaking out between itself and the U.K., where Alaska was expected to be seized. Driven by a power-political consideration rather than economic gains, Russia saw a way to secure some finances from the region by selling it to the U.S., who viewed its forests, animals, and geographical location as strategically beneficial (Hough, 2013:38).

24 At its peak in 1988, two million barrels/day were produced. Current production is equivalent of 600,000 barrels/day. Yet, this declining figure is put in perspective when compared to the Norwegian oil field Goliat, which produced about 100,000 barrels/day (Alaska Oil and Gas Association, 2017).

25 In the hydrocarbon-rich Beaufort Sea, there is an unresolved disagreement with Canada over the territorial boundary, and there is also political friction noticed in the way Russia and Canada both claim the Northwest passage and the Northern Sea Route as internal waters. The U.S. contests this and views these claims as inconsistent with international law, and rather argues the straits to be transit passages – thus making it free to sail, or to follow the coastal state's national legislation (U.S. Department of Defense, 2019).

26  This would also be telling of the U.S. to use public meetings as staged and theatrical acts on performance.

## References

Abedi, M. (2019). *Canada Positioned Itself as a World Leader on Climate Change – Is It?* Global News Canada, September 26, 2019.

Alaska Oil and Gas Association. (2017). *Facts and Figures.* Accessed on May 17, 2017, from www.aoga.org/facts-and-figures.

Alaska Rescue Coordination Center. (2017). *Search and Rescue Contacts.* Accessed on December 22, 2017, from https://sarcontacts.info/contacts/alaska-rescue-coordination-center/

Aliyev, N. (2019). *Russia's Military Capabilities in the Arctic,* International Centre for Defense and Security, June 25, 2019. Accessed from https://icds.ee/russias-military-capabilities-in-the-arctic/.

Arctic Council Ministerial Meeting. (2000). *Notes from the Second Ministerial Meeting.* 2nd Ministerial Meeting. Barrow, Alaska, U.S.A., October 12–13, 2000.

Astrasheuskaya, N. (2019). "US and EU Sanctions Take Toll on Russian Oil and Gas Exploration." *Financial Times,* November 11, 2019.

Bachman, J. (2010). "Special Report: Oil and Ice: Worse than the Gulf Spill?" Reuters, November 8, 2010.

Barrera, J. (2011). "While Harper Talked tough with NATO on Arctic, U.S. Believed PM All Bark No Bite." *APT National News,* May 11, 2011.

Bergh, K. and Klimenko, E. (2016). "Understanding National Approaches to Security in the Arctic," in Jakobson, L. and Melwin, N. (eds.). *The New Arctic Governance.* SIPRI (Stockholm International Peace Research Institute) Research Report No. 25. Oxford: Oxford University Press.

Brait, E. (2015). "Portland's Bridge-Hangers and 'Kayaktivists" Claim Win in Shell Protest." *The Guardian,* July 31, 2015.

Brende, B. (2017). *Panel discussion: The Arctic in a Global Context.* The Arctic Frontiers Conference, Tromsø, Norway. January 24, 2017, published online January 25, 2017. Accessed from www.youtube.com/watch?v=AWx2yzWm77Y

——— (2015). "The Arctic. Important for Norway, Important for the World," *Harvard International Review,* April 16, 2015.

British Broadcasting Corporation (BBC). (2017). *Russia's New Arctic Trefoil Military Base Unveiled with Virtual Tour.* April 18, 2017.

Burke, D. C. (2017). "Leading by Example: Canada and its Arctic stewardship role," *Int. J. Public Policy,* Vol. 13: 1/2, pp. 36–52.

Byers, M. (2013). *International Law and the Arctic. Cambridge Studies in International law and Comparative Law.* Cambridge: Cambridge University Press.

Campbell, L. M., Corson, C., Gray, N. J., MacDonald K. I. and Brosius, P. J. (2014). "Studying Global Environmental Meetings to Understand Global Environmental Governance: Collaborative Event Ethnography at the Tenth Conference of the Parties to the Convention on Biological Diversity," *Global Environmental Politics,* Vol. 14:3, pp. 1–20.

Canada: Arctic Council Ministerial Statement. (2017). *The Honourable Chrystia Freeland, Canada's Minister of Foreign Affairs.* 10th Ministerial Meeting. Fairbanks, Alaska, May 11, 2017. Accessed via oaarchive.arctic-council.org.

Canada's Arctic Policy/Strategy. (2019). *Canada's Arctic and Northern Policy Framework.* Ottawa: Government of Canada.

———— (2009). *Canada's Northern Strategy: Our North, Our Heritage, Our Future.* Published under the authority of the Minister of Indian Affairs and Northern Development and Federal Interlocutor for Métis and Non-Status Indians. Government of Canada. Ottawa: Government of Canada.

Carlsson L. (2019). "Banbrytande isbrytare på gång i Sverige." *Radio Sweden, Science Radio,* January 10, 2019.

Carlsson, M. and Winnerstig, M. (2016). *Irreconcilable Differences. Analyzing the Deteriorating Russian-US Relations.* May 2016, FOI. Report no FOI-R – 4276 – SE.

CBC News. (2010). *Arctic Sovereignty a Priority: Harper.* CBC News – Canadian Broadcasting Corporation. August 23, 2010.

———— (2009). *Canada against EU Entry to Arctic Council because of Seal Trade Ban.* CBC News – Canadian Broadcasting Corporation. April 29, 2009.

CCPI (2020). *The Climate Change Performance Index 2020.* Burck, J., Hagen, U., Höhne, N., Nascimento, L. and Bals C. Germanwatch and Climate Action Network Europe (CAN Europe), December 10, 2019.

Centre for High North Logistics. (2020). *CHNL Information Office: Transit statistics.* Accessed on June 12, 2017, from www.arctic-lio.com/nsr_transits.

Chater, A. (2016). "Explaining Russia's Relationship with the Arctic Council," *International Organizations Research Journal,* Vol. 11:4, pp. 41–54.

Conley, H. A. and Melino, M. (2019). *The Implications of U.S. Policy Stagnation toward the Arctic Region.* CSIS (Center for Strategic and International Studies). May 2019.

Conley, H. A. and Rohloff, C. (2015). *The New Ice Curtain. Russia's Strategic Reach to the Arctic.* CSIS (Center for Strategic and International Studies). August 2015.

David, M. (2013). *U.S. National Strategy for the Arctic Region: Strong Foothold or Thin Ice?* The Arctic Institute, Centre for Circumpolar Security Studies. May 13, 2013.

Death, C. (2011). "Summit Theatre: Exemplary Governmentality and Environmental Diplomacy in Johannesburg and Copenhagen," *Environmental Politics,* Vol. 20:1, pp. 1–19.

Defense Command Denmark. (2017). *Joint Arctic Command.* Last updated November 8, 2017, accessed on November 22, 2017, from www2.forsvaret.dk/eng/Organisation/ArcticCommand/Pages/ArcticCommand.

Denmark: Arctic Council Ministerial Statement. (2019). *Minister for Foreign Affairs of the Kingdom of Denmark, Mr. Anders Samuelsen.* 11th Ministerial meeting. Rovaniemi, Finland, May 7, 2019. Accessed via oaarchive.arctic-council.org.

———— (2013). *Ministerial Meeting in the Arctic Council – Intervention by the Danish Minister for Foreign Affairs, Villy Søvndal.* 8th Ministerial Meeting. Kiruna, Sweden, May 15, 2013.

———— (2009). *Speech by Denmark's Minister of Foreign Affairs.* 6th Ministerial meeting. Tromsø, Norway, April 29, 2009.

Denmark's Arctic Strategy. (2011). *Denmark, Greenland and the Faroe Islands. Kingdom of Denmark: Strategy for the Arctic 2011–2020.* Copenhagen, Nuuk, Tórshavn: Ministry of Foreign Affairs.

Dodds, K. and Ingimundarson, V. (2012). "Territorial Nationalism and Arctic Geopolitics: Iceland as an Arctic Coastal State," *The Polar Journal,* Vol. 2:1, pp. 21–37.

Earthjustice. (2013). *Petition to the Inter-American Commission on Human Rights Seeking Relief from Violations of the Right of Arctic Athabaskan Peoples*

*Resulting from Rapid Arctic Warming and Melting Caused by Emissions of Black Carbon by Canada*. Summary. Submitted April 23, 2013, on behalf of Arctic Athabaskan Council.

Equinor. (2020). *Sustainability & Climate*. Accessed on February 4, 2020, from www.statoil.com/en/how-and-why/climate-change.html.

EUR-Lex (2009). *EU Law*. Regulation No. 1007/2009 of the European Parliament and of the Council of September 16, 2009 on Trade in Seal Products.

European Commission. (2017a). *Countries and Regions: Iceland*. Last updated December 15, 2017, accessed on December 22, 2017, from http://ec.europa.eu/trade/policy/countries-and-regions/countries/iceland/.

——— (2017b). *Countries and Regions; Norway*. Last updated February 22, 2017, accessed on May 4, 2017, from http://ec.europa.eu/trade/policy/countries-and-regions/countries/norway/.

Fears, D. and Eilperin, J. (2016). "President Obama Bans Oil Drilling in Large Areas of Atlantic and Arctic Oceans." *The Washington Post*, December 20, 2016.

Finkler, H. and Kadas, R. (2016). "The 20th Anniversary of the Ottawa Declaration. Interview with Harald Finkler and Robert Kadas," in Heininen, L., Exner-Pirot, H. and Plouffe, J. (eds.), *Arctic Yearbook 2016. The Arctic Council: Twenty Years of Regional Cooperation and Policy-Shaping*. Iceland: Northern Research Forum.

Finland: Arctic Council Ministerial Statement. (2019). *Speech by Minister Timo Soini*. 11th Ministerial Meeting. Rovaniemi, Finland, May 7, 2019. Accessed via oaarchive.arctic-council.org.

——— (2017). *Statement by Minister Timo Soini*. 10th Ministerial Meeting. Fairbanks, Alaska, May 11, 2017.

——— (2015). *Statement by Mr. Erkki Tuomioja, Minister for Foreign Affairs of Finland*. 9th Ministerial Meeting. Iqaluit, Canada, April 24, 2015.

——— (2013). *Statement by Mr. Erkki Tuomioja, Minister for Foreign Affairs of Finland*. 8th Ministerial Meeting. Kiruna, Sweden, May 12, 2013.

——— (2002). *Opening Speech by Mr. Erkki Tuomioja, Minister for Foreign Affairs of Finland*. 3rd Ministerial Meeting. Inari, Finland, October 9–10, 2002.

Finland's Arctic Strategy. (2013). *Finland's Strategy for the Arctic Region 2013*. Government resolution on August 23, 2013. Prime Minister's Office Publications 16/2013. Helsinki: Prime Minister's Office.

——— (2016) *Update to the Arctic Strategy 2016*. The Government's strategy session September 26, 2016. Prime Minister's Office.

Finnemore, M. (1996). *National Interests in International Society*. Ithaca: Cornell University Press.

Fogh Rasmussen, A. (2015). "A Place Apart. A Peaceful Arctic No More?" *Harvard International Review,* Spring 2015, pp. 45–48.

Geertz, C. (1980). *Negara. The Theatre State in Nineteenth-Century Bali*. Princeton: Princeton University Press.

Government of Canada. (2016). *Modification*. Address by Parliamentary Secretary Pamela Goldsmith-Jones on behalf of Minister Dión, marking the 20th anniversary of the Arctic Council, September 29, 2016, Ottawa, Ontari.

Government of Norway. (2015a). "Norway Support Efforts to Eliminate Gas Flaring by 2030." News, April 17, 2015. Accessed from www.regjeringen.no/en/aktuelt/eliminate_flaring/id2407055/.

———— (2015b) *Oppdatering av forvaltningsplanen for Barentshavet – Lofoten.* News, published April 25, 2015. Accessed from www.regjeringen.no/no/aktuelt/ oppdatering-av-forvaltningsplanen-for-barentshavet – lofoten/id2408548/.

Granholm, N. (2016). *Arktis under förändring – standarbilden utmanas.* Swedish Defence Research Agency, Report.

Harding, L. (2010). "Vladimir Putin calls for Arctic claims to be resolved under UN law." *The Guardian*, September 23, 2010.

Hilde, P. S. (2014). "Armed Forces and Security Challenges in the Arctic," in Tamnes, R. and Offerdal, K. (eds.). *Geopolitics and Security in the Arctic. Regional Dynamics in a Global World.* London and New York: Routledge.

Hønneland, G. (2016). *Russia and the Arctic. Environment, Identity and Foreign Policy.* London and New York: I.B. Tauris.

Hough, P. (2013). *International Politics of the Arctic: Coming in from the Cold.* London and New York: Routledge.

Huebert, R. (2011). "Submarines, Oil Tankers, and Icebreakers: Trying to Understand Canadian Arctic Sovereignty and Security," *International Journal*, Vol. 66:4, pp. 809–824.

Humpert, M. (2019). "U.S. Coast Guard Awards Contract for New Polar Ice Class Icebreaker." *High North News,* April 30, 2019.

Iceland: Arctic Council Ministerial Statement. (2019). *Presentation of the Icelandic Chairmanship Program, Mr. Gudlaugur Thór Thórdarson, Minister for Foreign Affairs.* 11th Ministerial meeting. Rovaniemi, Finland, May 7, 2019. Accessed via oaarchive.arctic-council.org.

———— (2006). *Address by H.E. Valgerður Sverrisdóttir. Minister for Foreign Affairs of Iceland.* 5th Ministerial Meeting. Salekhard, Russia, October 26, 2006.

Iceland's Arctic Strategy. (2011). *A Parliamentary Resolution on Iceland's Arctic Policy.* Approved by Althingi at the 139th legislative session, March 28, 2011.

*Iceland's Chairmanship Program for the Arctic Council 2019–2021.* Available via arctic-council.org.

Interviews (2014–2015). Interviews conducted between December 2014 and December 2015 with state representatives and other representatives active in Arctic Council cooperation and negotiation.

Jacobsen, M. (2016). *Denmark's Strategic Interests in the Arctic: It's the Greenlandic Connection, Stupid!* The Arctic Institute, Centre for Circumpolar Security Studies. May 4, 2016.

Jensen, L. C. (2012). "Norwegian Petroleum Extraction in Arctic Waters to Save the Environment: Introducing Discourse Co-optation as a New Analytical Term," *Critical Discourse Studies,* Vol. 9:1, pp. 29–38.

Lind, G. (2011). "Arktisambassadören: Det svenska "flaggskeppet" fick ett entusiastiskt mottagande." Published in *Aktuell Hållbarhet*, November 18, 2011.

Kerry, J. (2015). *Remarks at the Presentation of the U.S. Chairmanship Program at the Arctic Council Ministerial.* 9th Ministerial Meeting. Iqaluit, Canada. April 24, 2015.

Luhn, A. (2016). "The Town that Reveals How Russia Spills Two Deepwater Horizons of Oil Each Year." *The Guardian*, August 5, 2016.

Lunn, S. (2015). *Leona Aglukkaq Says Letter to Provinces on Emissions Targets Wasn't an Attack.* CBC News – Canadian Broadcasting Corporation. April 17, 2015.

Majlard, J. (2016). "Polarforskning kräver efterträdare till Oden." *Svenska Dagbladet – SvD.* April 6, 2016.

Ministry for Foreign Affairs. (2016). *Finland Promotes Environment, Stability, Vitality and Viability in the Arctic Region.* Press release 419/2016, October 5, 2016. Government Communications Department.

National Energy Authority of Iceland. (2017). *Oil and Gas Exploration.* Accessed on July 27, 2017, from www.nea.is/oil-and-gas-exploration/.

Nicol, H. N and Heininen, L. (2014). "Human Security, the Arctic Council and Climate Change: Competition or Co-existence?" *Polar Record,* Vol 50:1, pp. 80–85.

Nilsen, T. (2019). "Russia Removes Critical Voices Ahead of Arctic Council Chairmanship, Claims Indigenous Peoples Expert." *The Independent Barents Observer*, November 27, 2019.

Nilsson, A. E. (2007). "A Changing Arctic Climate. Science and Policy in the Arctic Climate Impact Assessment." Dissertation. Linköping Studies in Art and Science No 386. Department of Water and Environmental Studies, Linköping University. Linköping: UniTryck.

Nord, D. C. (2016a). *The Arctic Council: Governance within the Far North.* New York: Routledge.

——— (2016b). *The Changing Arctic. Consensus Building and Governance within the Arctic Council.* New York: Palgrave Macmillan US.

Norway: Arctic Council Ministerial Statement. (2015). *Arktisk Råds ministermøte.* 9th Ministerial Meeting. Iqaluit, Canada, April 24, 2015. Accessed via oaarchive. arctic-council.org.

——— (2013). *National Statement by Norwegian Minister of Foreign Affairs, Espen Barth Eide.* 8th Ministerial Meeting. Kiruna, Sweden, May 15, 2013.

——— (2009). *Norwegian MFA Jonas Gahr Støres welcoming remarks at the sixth Ministerial Meeting of the Arctic Council.* Tromsø, Norway, April 29, 2009.

——— (2006a). *Minister of Foreign Affairs Jonas Gahr Støre.* 5th Ministerial Meeting. Salekhard, Russia, October 26, 2006.

——— (2006b). *Concluding statement by Minister of Foreign Affairs Jonas Gahr Støre.* 5th Ministerial Meeting. Salekhard, Russia, October 26, 2006.

Norway's Arctic Strategy. (2017). *Norway's Arctic Strategy. Between Geopolitics and Social Development.* Ministry of Foreign Affairs, and Ministry of Local government and Modernization.

——— (2014). *Norway's Arctic Policy. Creating Value, Managing Resources, Confronting Climate Change and Fostering Knowledge. Developments in the Arctic Concern Us All.* Norwegian Ministry of Foreign Affairs.

——— (2006). *The Norwegian Government's High North Strategy.* Ministry of Foreign Affairs.

Offerdal, K. (2014). "Interstate Relations: The Complexities of Arctic Politics," in Tamnes, R. and Offerdal, K. (eds.). *Geopolitics and Security in the Arctic. Regional Dynamics in a Global World.* London and New York: Routledge.

Oseland, K. M. and Raspotnik, A. (2018). "The Oslo Arctic Dream." *About Energy,* December 18, 2018.

Pettersen, T. (2016). "Northern Fleet Gets Own Air Force, Air Defense Forces." *The Independent Barents Observer,* February 1, 2016.

Rahbek-Clemmensen, J. (2017). "The Ukraine Crisis Moves North. Is Arctic Conflict Spill-Over Driven by Material Interests?" *Polar Record,* Vol. 53:1, pp. 1–15.

——— (2016). "An Arctic Great Power? Recent Developments in Danish Arctic Policy," in Heininen, L., Exner-Pirot, H. and Plouffe, J. (eds.). *Arctic Yearbook*

2016. *The Arctic Council: Twenty Years of Regional Cooperation and Policy-Shaping.* Iceland: Northern Research Forum.

Restino, C. (2015). "Opinion: Federal Lawmakers Must Tackle America's Arctic Identity Crisis." *The Bristol Bay Times,* August 28, 2015.

Reuters News Agency. (2017). "Donald Trump Signs Executive Order Aimed at Lifting Bans on Arctic Drilling," *The Telegraph,* April 28, 2017.

Russia: Arctic Council Ministerial Statement. (2013.) *Speech of Russian Foreign Minister Sergey Lavrov at the Eight Ministerial Session of the Arctic Council.* Kiruna, Sweden, May 15, 2013. Accessed via oaarchive.arctic-council.org.

———— (2006). *Address by Minister of Foreign Affairs of the Russian Federation Sergey Lavrov at the Fifth Ministerial Session of the Arctic Council.* Salekhard, Russia, October 26, 2006.

Russia's Arctic Strategy. (2013). *The Development Strategy of the Arctic Zone for the Russian Federation.* Approved by President V. Putin.

———— (2008). *Russian Federation Policy for the Arctic to 2020.* Adopted by President D. Medvedev on September 18, 2008.

SAO Discussion Paper. (2016a). *Discussion Paper on the Arctic Council's Work on Oil and Gas Issues.* Submitted by United States of America. Arctic Council SAO plenary meeting, Fairbanks, Alaska, March 16–17, 2016. Accessed from the Arctic Council Open Access Repository: https://oaarchive.arctic-council.org/handle/11374/1730.

———— (2016b). *Discussion Paper on Climate Change Activity in the Arctic Council.* Submitted by United States of America. Arctic Council SAO plenary meeting, Fairbanks, Alaska, March 16–17, 2016. Accessed from the Arctic Council Open Access Repository: https://oaarchive.arctic-council.org/handle/11374/1728.

SAO Meeting Minutes. (2019). *Arctic Council Senior Arctic Officials Meeting Hveragerði, Iceland.* November, 20–21, 2019. Report, pp. 1–21. Accessed via oaarchive.arctic-council.org.

———— (2018). *Levi, Finland.* March, 22–23, 2017. Report, pp. 1–19.

———— (2018). *Rovaniemi, Finland.* November, 1–2, 2018. Report, pp. 1–25.

———— (2017). *Juneau, Alaska, U.S.A.* March 8–9, 2017. Summary report, pp. 1–23.

———— (2017). *Oulu, Finland.* October, 25–26, 2017. Report, pp. 1–13.

———— (2016). *Fairbanks, Alaska, U.S.A.* March 16–17, 2016. Report, pp. 1–27.

———— (2015). *Whitehorse, Canada.* March 4–5, 2015. Final report, pp. 1–21.

———— (2015). *Anchorage, Alaska, U.S.A.* October 21–22, 2015. Summary report, pp. 1–24.

———— (2014a). *Yellowknife, Canada.* March 26–27, 2014. Final report, pp. 1–16.

———— (2014b). *Yellowknife, Canada.* October 22–23, 2014. Final draft report, pp. 1–19.

———— (2013). *Stockholm, Sweden.* March 20–21, 2013. Final report, pp.1–9.

———— (2013). *Yukon, Canada.* October 22–23, 2013. Final report, pp. 1–17.

———— (2012). *Stockholm, Sweden.* March 28–29, 2012. Final report, pp.1–6.

———— (2012). *Haparanda, Sweden.* November 14–15, 2012. Final report, pp. 1–11.

———— (2011). *Copenhagen, Denmark.* March 16–17, 2011. Final report, pp. 1–13.

———— (2011). *Luleå, Sweden.* November 8–9, 2011. Final report, pp.1–13.

———— (2010). *Ilulissat, Greenland.* April 28–29, 2010. Final report, pp. 1–23.

———— (2010). *Torshavn, Faroe Islands.* October 19–20, 2010. Final report, pp. 1–17.

———— (2009a). *Copenhagen, Denmark.* February 10, 2009. Final report, pp. 1–4.

———— (2009b). *Copenhagen, Denmark.* November 12–13, 2009. Final report, pp. 1–15.

—— (2008). *Svolvær, Norway.* April 23–24, 2008. Final report, pp. 1–17.

—— (2008). *Kautokeino, Norway.* November 19–20, 2008. Final report, pp.1–16.

—— (2007). *Tromsø, Norway.* April 12–13, 2007. Draft minutes, pp. 1–12.

—— (2007). *Meeting Narvik, Norway.* November 28–29, 2007. Final report, pp. 1–21.

—— (2005). *Yakutsk, Russia.* April 6–7, 2005. Draft minutes, pp. 1–15.

—— (2005). *Khanty-Mansiysk, Russia.* October 12–14, 2005. Draft minutes, pp. 1–19.

—— (2003). *Reykjavik, Iceland.* April 9–10, 2003. Minutes, pp. 1–17.

—— (2003). *Svartsengi, Iceland.* October 23–24, 2003. Draft minutes, pp. 1–15.

—— (2002). *Oulu, Finland.* May 15–16, 2002. Minutes, pp. 1–22.

—— (2002). *Inari, Finland.* October 7–8, 2002. Minutes, pp. 1–15.

—— (2001). *Rovaniemi, Finland.* June 12–13, 2001. Minutes, pp. 1–13.

—— (2001). *Espoo, Finland.* November 6–7, 2001. Minutes, pp. 1–24.

—— (2000). *Fairbanks, Alaska, U.S.A.* April 27–28, 2000. Minutes (revised – 10/03/00), pp. 1–37.

—— (1999). *Washington, D.C., U.S.A.* November 18–19, 1999. Minutes (draft – 4/12/00), pp. 1–22.

—— (1999). *Anchorage, Alaska, U.S.A.* May 5–6, 1999. Minutes (revised – 28/09/99), pp. 1–17.

Sergunin, A. and Konyshev, V. (2016). *Russia in the Arctic. Hard or Soft Power?* Series of Soviet and Post-Soviet Politics and Society, Vol. 149, Umland, A. (ed.). Stuttgart: Ibidem-verlag.

Solberg, E. (2017). *Speech by Prime Minister Erna Solberg at the Arctic Frontiers Conference.* Accessed January 23, 2017, from www.regjeringen.no/en/aktuelt/arctic-frontiers/id2528099/.

Staalesen, S. (2016). "Government Says New Monster Ice Breaker Will Become Reality." December 16, 2016.

—— (2015). "Russian Military Builds Four More Arctic Bases." *The Independent Barents Observer,* October 23, 2015.

Staalesen, S. and Nilsen, T. (2012). "Moscow Orders Closure of Indigenous People's Organization," *The Independent Barents Observer,* November 12, 2012.

Stokke, O. S. (2014). "International Environmental Governance and Arctic Security," in Tamnes, R. and Offerdal, K. (eds.). *Geopolitics and Security in the Arctic. Regional Dynamics in a Global World.* London and New York: Routledge.

Stortinget (2016). *Innstilling fra energi- og miljøkomiteen om Samtykke til ratifikasjon av Paris-avtalen av 12. desember 2015 under FNs rammekonvensjon om klimaendring av 9. mai 1992.* Inst. 407. S. Prop. 115 S (2015–2016). Energi- of Miljøkomiteen.

Sweden: Arctic Council Ministerial Statement. (2017). *Speech by Minister for Foreign Affairs Margot Wallström.* 10th Ministerial meeting. Fairbanks, Alaska, May 11, 2017. Accessed via oaarchive.arctic-council.org.

—— (2015). *Statement by Sweden.* 9th Ministerial Meeting. Iqaluit, Canada, April 24, 2015.

—— (2013). *Statement by Sweden.* 8th Ministerial Meeting. Kiruna, Sweden, May 15, 2013.

—— (2011). *Presentation by the Minister for foreign Affairs of Sweden, Mr. Carl Bildt, on the Swedish Programme for the Chairmanship of the Arctic Council.* 7th Ministerial Meeting. Nuuk, Greenland, May 12, 2011.

———— (2009). *Remarks by the Minister for Foreign Affairs of Sweden, Carl Bildt.* 6th Ministerial Meeting. Tromsø, Norway, April 29, 2009]

Sweden's Arctic Strategy. (2016). *New Swedish Environmental Policy for the Arctic.* Ministry of the Environment and Energy.

———— (2011). *Sweden's Strategy for the Arctic Region.* Stockholm: Ministry for Foreign Affairs.

Swedish Maritime Administration (SMA). (2012). *Sjöfartsverkets treårsplan, 2013– 2015.* March 1, 2012.

Swedish Polar Research Secretariat. (2016). *Utredning om hur statens behov av ett forskningsfartyg med isbrytande kapacitet avsett för vetenskapliga expeditioner bäst ska kunna tillgodoses.* Slutrapport, dnr 2016-74, December 30, 2016. Stockholm: Polarforskningssekretariatet.

Taksøe-Jensen, P. (2016). *Danish Defense and Diplomacy in Times of Change.* A review of Denmark's Foreign and Security Policy. Executive Summary. Ministry of Foreign Affairs of Denmark. May 2016.

U.S.: Arctic Council Ministerial Statement. (2019) *Statement by Secretary of State Michael R. Pompeo.* 11th Ministerial Meeting. Rovaniemi, Finland, May 7, 2019. Accessed via oaarchive.arctic-council.org.

———— (2015) *Remarks at the Presentation of the U.S. Chairmanship Program at the Arctic Council Ministerial.* 9th Ministerial Meeting. Iqaluit, Canada, April 24, 2015.

U.S. Arctic Strategy. (2013). *National Strategy for the Arctic Region.* Approved by President B. Obama, May 2013.

———— (2009). *National Security Presidential Directive and Homeland Security Presidential Directive.* Approved by President G. W. Bush January 9, 2009.

U.S.-Canada Joint Arctic Leaders' Statement. (2016). *The White House: Office of the Press Secretary.* Accessed December 20, 2016, from https://obamawhitehouse. archives.gov.

U.S. Department of Defense. (2019). *U.S. Department of Defense Arctic Strategy.* Report to Congress, June 2019.

The White House. (2015). *President Obama's Trip to Alaska.* Accessed on June 14, 2016, from https://obamawhitehouse.archives.gov.

Zagorski, A. (2014). "Russia's Arctic Governance Policies," in Jakobson, L. and Melwin, N. (eds.). *The New Arctic Governance.* SIPRI Research Report No. 25. Oxford University Press.

# 5 When Sovereignty Is Expected to Interfere: A Micro-Level Departure in Negotiations on Marine Oil Spill Prevention

## Oil Spill Prevention in the Arctic

The theme of this chapter is oil pollution, or rather *prevention* of oil spills. The specific empirical focus is the negotiation process carried out by the Arctic Council Task Force on Arctic Marine Oil Pollution Prevention (TFOPP). By firstly investigating arguing among the eight Arctic states, and secondly their role performances, the chapter offers a micro-level perspective on consensus-reaching, and not least, the difficulties that accompany such a process. Ultimately the chapter reveals, and illustrates, social interaction to potentially encompass role learning.

For five meetings in 2014, the Arctic states – with input provided from national experts and advisors within coast guards, industries, and shipping, as well as from indigenous representatives – met to develop an "action plan or other arrangement on oil pollution prevention" (*Kiruna Declaration*, 2013). In a context of rapid Arctic warming and melting sea ice, increasingly exposing potentially offshore drilling opportunities, a TFOPP participant representing the indigenous people explained what an underlying logic and motivation for Arctic cooperation on oil spill prevention could be:

> We are in a little bit of danger. There is less focus now on oil and gas because the prices have fallen so much. Exploration has fallen, particularly in the marine environment, which is tremendously costly. Probably no one thinks that is going to continue over the long run but expects increased activity once again. So now would be the time to get a good handle on prevention in the Arctic.
>
> (Interview 2015)

The Arctic states agreed to a non-legal agreement titled 'Framework Plan for Cooperation on Prevention of Oil Pollution from Petroleum and Maritime Activities in the Marine Areas of the Arctic' (hereinafter Framework plan), and it states preventive measures for Arctic petroleum operators and maritime transporters to consider. As described in hindsight, the general feeling amongst the task force participants would be that of a negotiating process that started off with quite high expectations, but where ambitions lowered as interaction went on.[1]

The chapter has the following ambition: to reveal how social interaction, like negotiations, is making states aware of who they are and what they want; to show how the act of arguing (i.e., negotiating) sometimes leads the way toward common understanding, and sometimes astray; and to provide input as regards others' expectations and the behavioral impact that may follow. Enhanced knowledge on cooperative/environmental resistance is deemed important should environmental governance be enhanced.

The applied micro perspective does furthermore aim to illustrate role-playing as a social process at least partly informed by cooperative cues and demands. The analysis will be structured according to the following parameters of importance whenever actors aim for consensus-reaching: (1) common understanding; (2) arguing; and (3) role learning. Overall, this chapter aims to show that expectations constitute a formative dimension of state behavior, which implies the behavior of states to potentially be molded – learned – through cooperative processes. However, prior to the investigation of TFOPP, a background on the issue of oil pollution within the AC will be provided, serving as a contextual background to those discussions held in TFOPP.

### Oil Pollution and AC Cooperation on Its Prevention: A Background

*Oil and Polar Bears – Arctic Symbols*

Oil and the Arctic have since long an established bond. Russia struck oil already in 1915 and started its onshore development in the 1930s. A significant intensification – now also with gas discoveries – occurred in the 1960s throughout the 1980s (Kontorovich, 2015). In the U.S., the breakthrough discovery of Prudhoe occurred in the 1960s, and Canada, who had discovered oil in its Northwest territories already in 1920, went offshore in the Beaufort Sea in the early 1970s (Hough, 2013:21). By the 1980s, all current hydrocarbon developers in the Arctic, such as Canada, Norway, Russia, and the U.S., had gone offshore in their search for oil and gas,[2] and made discoveries.

Throughout history, natural resources have represented an economic gateway to the Arctic, as closely related to Arctic opportunities as polar bears have popularly been depicted as symbols of what is now threatened: Arctic wilderness, nature's strength, and endless icy landscapes. Both polar bears and the Arctic energy industry represent two giants of the Arctic, although one more symbolically than the other – the former as representing the untouched nature, and the latter for economic reasons. Thus, if polar bears in the popular discourse is the Arctic, then it would make practical sense that the Arctic governance regime of the 1970s, when the first oil boom struck the region, included a 1973 agreement on Conservation of polar bears but everything else was practically left environmentally unregulated.

International agreements and conventions on oil pollution, however, were quick to follow on oil operating activities and, for instance, concerned civil liability for oil pollution damage (1969) and marine pollution through waste dumping as well as land-based sources (1972/1974). In 1990, the IMO adopted international regulations calling for ships and offshore units to have oil pollution emergency plans, report incidents, and stockpile emergency equipment (International Convention on Oil Pollution Preparedness, Response and Cooperation, 1990). In addition, the UNCLOS (1982) set out to establish the jurisdictional basis for maritime activities, including limits for territorial sea, transit passages, and EEZs. The convention is not Arctic specific, but Article 234 deals with ice-covered areas and provides upgraded sovereignty to coastal states to develop and administer maritime regulations of choice, given these types of risky waters. This is not least considered important by Canada and Russia, with their long coastal lines and exposure to transit passages. Lastly, requirements on ship construction and equipment for transporting dangerous goods were introduced through the MARPOL (International Convention for the Prevention of Pollution from Ships) convention (1973/1978), requirements that more recently have been further elaborated through the *Polar Code* (2014).

Early conventions were mostly dedicated to a prohibition of careless environmental as well as safety behavior. Over the years, the discourse on oil pollution has been broadened to include variables pertaining to the question of 'if': what to do *if* an accident occurs, what to do *if* oil is spilled? Whereas polar bears today often symbolize a threatened future due to climate change,[3] they have also portrayed a more specific threat represented by oil extraction and transportation. When the EPPR in 1998 presented its first field guide on how to respond to oil spills in Arctic waters, the cover picture was that of a great polar bear. Indeed, the fact – and imagery – of polar bears grooming in oil is an image no one wants to have caused. Yet, the vulnerability of the Arctic environment toward an oil spill accident is, of course, more extensive than polar bears. Owing to remoteness and icy conditions that would stall rescue operations and clean-up, the consequences here would be more severe than accidents elsewhere in the (PAME, 2014; WWF, 2014). Symbols are nonetheless important since they appeal to understandings of a certain phenomenon, and what to consider appropriate behavior. Within the Arctic Council, discussions have reached a point where preparedness and response are not enough: the question of what to do if an accident occurs has thus been replaced, or at least complemented, by the even more demanding question of what to do to *prevent* oil accidents?

*Oil Pollution and the Arctic Council – Prior to Expecting a Twenty-First Century 'Arctic Boom'*

When Arctic oil pollution cooperation begun in 1991, through the AEPS, the international legal framework covering the issue was described as insufficient. The Arctic states pointed toward the Arctic as a special case within pollution regulation: the Arctic would be one of the areas mostly vulnerable to oil pollution, where low temperatures, darkness, and cold were described to lead to slower decomposition of oil compared to tempered regions, and where oil on feathers and fur made animals lose their insulation properties, in addition to being toxic if ingested. Transportation, especially on rivers, oil production, and to a lesser extent, oil exploration, were considered the greatest risk activities (AEPS, 1991:14–15).

The concern from the Arctic states came partly from experience. In 1989, the oil tanker Exxon Valdéz left the Prudhoe Bay oil field fully loaded. Soon came the ship radio call: "Yeah, this is Valdez. [...] We've fetched up, hard aground, north of Goose Island off Bligh Reef and ... evidently ... leaking some oil..." (National Ocean Service, 2014). Eleven million gallons of crude oil were spilled into the sea, damaging twenty-eight species of animals and plants in what up until then would count as the worst oil spill catastrophe in the U.S. history, as well as the Arctic. It would take more than ten years until the first species had recovered, whereas others – after several decades – still have not experienced recovery (ibid.). When the Arctic states started to cooperate on oil pollution, through AEPS, a common understanding of the need thereof was though not only perceived but also experienced. The working group EPPR was established to facilitate cooperation and information sharing amongst the Arctic states and directed responses to environmental emergencies.

In one of its first reports, an environmental risk analysis report, EPPR found transportation and storage of oil as the greatest Arctic threats. The probability of an oil spill was here assessed as high, with greatest risks found in Russia in relation to both oil production and exploration, and harboring and transport (EPPR, 1998:vi). The working group PAME also presented in 1997 – on a U.S. initiative – work on oil pollution through its first *Arctic Offshore Oil and Gas Guidelines*. These were guidelines intended for the regulator's community, those responsible for regulation of oil and gas activities, with the aim of developing a set of practices to be used in a consistent (harmonized) way across the Arctic states. Principles such as precautionary approach, polluter pays, and sustainable development should be the foundation of all oil and gas activities, so the guidelines read (PAME, 1997:6). The AC episteme, as well as states and other actors, were thus aware early on of the oil issues, but not yet ready to cooperate beyond their national jurisdictions.

To go back to Exxon Valdéz, this accident was an environmental catastrophe that put the spotlight on risks associated with Arctic oil activities. But it was also an accident that showed the industry responding in a responsible manner, at least this was how Arctic stakeholders with an interest of resource extraction understood it. The mayor of North Slope Borough, for instance, explained how the industry had learned from its mistakes, and how oil revenues contributed to improved life for the Far North (AC Ministerial meeting Barrow, 2000).[4] As such social and economic benefits generally were considered to outweigh environmental risks.

*Expecting the Strike of 'an Oil Boom'*

In early 2016, the U.S. presented that the Arctic Council had made some 238 recommendations on oil and gas activities (SAO discussion paper, 2016). Since oil pollution is a cross-cutting issue, one finds these recommendations spread across the AC reports, relating to such things as human development, shipping, and biodiversity assessments. The majority, however, is found in defined reports on oil and gas. As described by the U.S., in the early days of drafting reports and recommendations within Arctic cooperation, Arctic states had just entered commercial oil and gas activities. Some ten years later though one expected markedly increased activity levels within this field, both onshore and offshore (ibid.). In parallel to this development, recommendations have become more intensified and demanding in their formulations.

AMAP's *Assessment of Oil and Gas Activities in the Arctic* (2007) was the first report to go. "Pollution cannot be reduced to zero" the report said, "only be minimized," concluding that with more activity expected, an increasingly significant proportion of human hydrocarbon input to Arctic pollution would be expected as well (ibid.:v–vi). WGs all called for more cooperation and greater carefulness from operators.[5] Then, in 2010, the blowout of the Macondo oil well clearly revealed the dangers with offshore oil extraction, and what can happen when a pollution accident occurs. In addition to eleven causalities, it was considerably more extensive than the Exxon Valdéz, in terms of spill, leaking at least 130 million gallons of oil straight into the Gulf of Mexico (National Oceanic and Atmospheric Administration, 2017), compared to 11 million gallons in the Exxon Valdéz equivalent. The blowout caused President Obama to call it "the worst environmental disaster America has ever faced," an epidemic to be fought for years[6] (Obama, 2010). Yet, the exploding rig was 'luckily' based in the Mexican Gulf with more favorable conditions regarding clean-up, compared to colder waters. Here, an offshore accident could be much worse in terms of effects on people and ecosystems, as well as longer lasting (PAME, 2014).

The year after Macondo, in 2011, the Arctic Council mandated a task force, led by Norway, Russia, and the U.S., to negotiate how to enhance safety and security regarding marine oil pollution. The result was a binding *Agreement on Cooperation on Marine Oil Pollution Preparedness and Response in the Arctic* (MOSPA, 2013). The idea was to establish how to act should an oil spill occur, for instance, by notifying as well as assisting other states, and promoting information exchange by suggestively conducting joint exercises (ibid.). Although participants entered the negotiation process with low expectations when it came to reaching an agreement, due to 'a range of implicated players' (SAO meeting Luleå, 2011), the task force managed to reach a binding agreement and it was a process praised by the SAOs (SAO meeting Haparanda, 2012). The content of the agreement, however, has been criticized for providing little guidance on how all of this should be carried out in practice, what is required by states in terms of oil spill recovery – which technology or what regulator schemes they need to have put in place (Rottem, 2015:55; 2016:164–165). It is difficult to identify any operational consequences of the agreement, which "is more important as a symbol for Arctic cooperation than as a practical mechanism" (ibid., 2015:55).

Still, even a symbol (for Arctic cooperation) may push normative frontiers forward, and steer cooperation in a specific direction. To have the MOSPA agreement put in place was considered necessary in order to take the next step, toward oil pollution *prevention*. It was logical, since entering the more demanding task of prevention (compared to preparedness) required a confirmation of each other being dedicated to the cause (interview 150623a). When TFOPP was created, it was the third task force in a row dedicated to marine safety and security operations, after *Search and Rescue* (2011) and MOSPA (2013). Several TFOPP participants knew each other from these processes. As such, the stepwise negotiation process involved a search for elements of trust building, in relation to the cause but also on a more personal level.

Preventive measures would at that time for TFOPP be politically – as well as industrially – sanctioned and motivated, not least with Deepwater in mind. Notions on oil pollution prevention were that it is *salient* – that is, a relevant topic given the Arctic context as particularly vulnerable to oil spills.[7] Following Clark, Mitchell and Cash (2006) preventing oil spills would furthermore be a *legitimate* topic, given the 'empirical evidence' as represented by Deepwater Horizon. And lastly, it would be a *credible* topic given that emergency preparedness and response all had been dealt with for twenty years, adding trustworthy and useful knowledge on how to enhance safety and security regarding oil and gas activities. Consequently, it signals oil and gas activities as no longer fully protected under strict perceptions on sovereignty, but ready for cooperation.

## Negotiating Oil Spill Prevention: High Expectations, Lower Ambitions

The remaining part of this chapter is devoted to TFOPP and its consensus-reaching process on oil pollution prevention, with focus directed on how such consensus eventually came about. Such information is believed to shed important light on cooperative hurdles. The first part discusses the prevalence and characterization of common understanding and arguing within the process, concepts that here are believed to encapsulate change and resistance, respectively. The last part of the chapter illustrates, through a discussion of some states' role performances, how social interaction has actors to reflect upon who they are and what they want, given its context. As such both discussions pertain to learning as a social process.

At the first meeting of TFOPP, in mid-January 2014, the atmosphere was good, with great interest and a general feeling of something important about to happen (TFOPP Two-pager, Oslo, 2014). The successful history of the past two agreements as sprung task force negotiations made delegations reportedly arrive at the TFOPP process quite open minded. One Head of Delegation (HoD) explained it as a general feeling that prevailed, signaling "ok, we are making progress, the Arctic Council is getting stronger" (interview 150626). Four meetings later, when the process was to be concluded, the tone was another; now the co-chairs urged delegates to not be too caught up in particular details but be consensus-oriented toward a more concrete document (TFOPP Two-pager Helsinki, 2014). Compared to the beginning, TFOPP left behind a process with lowered ambitions and somewhat stalled enthusiasm: opinions on oil spill prevention as well as jurisdictional preferences had indeed been divergent. With the final document in hand, two things yet stood clear: firstly, all Arctic states had been committed to the task of reaching an agreement; and secondly, what to write – and read – into the document had been trickier.

### Common Understanding on the Need for Cooperation

#### Oil Spills as Arctic Threats

TFOPP and its different state delegations were united in two different ways: in a perception of accuracy – i.e., it being a valid claim – that oil spills constitute Arctic threats, and in the perception of cooperation as adding value to Arctic relations. To start with the former, one of the co-chairs, Ambassador (SAO) Vasiliev, has explained the underlying motivation for cooperation on the matter in the following way:

> Last year [2013] the first Arctic oil from the offshore ice-covered area was extracted. So actually there is no more discussion of whether we should produce Arctic oil or not – we are. Still, the Arctic is the Arctic, and there are so many challenges - climatic, technological, ice, etc. And the most important risk with oil production is the risk of a spill. Thanks to work done in the Arctic Council, we know what the consequences could be in the Arctic from an oil spill. There, it could be much more dangerous than in other parts of the world.
>
> A. Vasiliev, co-chair TFOPP, 2014[8]

In relation to the 'objective world,' as Habermas would have put it, the Arctic states shared a common understanding on oil spills as in need of prevention; science and experts had contributed with knowledge thereof. It was considered important to act and take safety measures 'now,' forestalling the future risk of spills. Interview accounts were telling of a need to act before the area "is crowded with activities" and to "agree before things happen" (interview 150506; 151221). This describes a cooperative logic derived from a wish to escape those economic interests that a more accessible Arctic region could imply, and that are expected to render a consensus significantly more difficult. Another delegate explains:

> Since conditions change rapidly in the Arctic, it is important to do something well in advance to regulate and protect what possibly can be protected. It is easier to make up rules now before any great economic activity will take place. There are currently large economic interests, but no economic activity. Therefore, now is the time to speed up the process drawing lines for activity.
>
> (Adviser 6)

The O&G industry was also present in TFOPP, and it was reported that several suggestions on preventive measures came from this group of representatives. However, several state representatives found that the industry seemed to prefer as little involvement as possible by the states. The reason, one business representative explained, was that most of the work already is coming from the industry: funding on projects, research, and standards. But the bulk of this work passed by largely unseen: the industry *is* engaged, this interviewee stated, rhetorically continuing, "think of all the interests the oil industry has of preventing oil pollution…" (interview 150324). Indeed, from an industrial perspective, there would be a strategic interest of minimizing accidents. Also, among state delegates, oil pollution prevention was reasoned about in a strategic way, where the Arctic states were described to be in the same boat should an accident occur:

> If a spill happens in the Arctic, no matter where, then it will affect the whole of the Arctic activity. Environmentalist groups will strongly call on a ban on drilling in the Arctic, no matter where the accident happens.
>
> (Negotiator 11)

The incentives for oil pollution prevention cooperation was thus laid bare and represented a common understanding of oil pollution being the critical issue for the Arctic future to come. The past had taught the Arctic states about the viable way forward:

> […] everybody there was very attuned toward what had happened in the Gulf 3-4 years earlier [the Macondo blowout in 2010]. And the understanding that such a thing could never happen in the Arctic, and the understanding of… you know, we all have different legal schemes to work under but we all have to work toward preventing that.
>
> (Interview 150521b)

When engaging in the TFOPP, any oil accident in the Arctic would mean a tremendous challenge – environmentally, socially, and economically, and on this the Arctic states all agreed.

*Acting as a Cooperator*

When the task force participants were asked to point out the most important factor for states coming to an agreement, a common answer would be that 'we were all dedicated to the task.' With this, they do not only refer to the prevention task per se, but perhaps above all the task of *cooperating*: to engage in and enhance cooperation seems to be derived from a common understanding of how the AC member states should behave. "In the AC one agrees," participants reasoned, and followed the same strategy as for MOSPA where "nothing would be agreed until everything was agreed" (SAO meeting Luleå, 2011). Equally, the establishment of the TFOPP originated from an approximately 25-year-old institutional AC structure always aiming forward, on to greater harmonization, exchange of experiences and data, joint exercises, and other things encouraging interaction based upon senses of commonness. One agreed, furthermore, because this was 'a want' expected to be shared by all. A HoD explains the underlying motivation for reaching an agreement in the TFOPP:

> Of crucial importance is that all eight Arctic states want, want to have cooperation. No state wants to leave the cooperation. Even Russia, who currently is very critical of much of the things carried out by the West, still wants to have this cooperation continuously. The same goes for Canada. We know that Russia foresees an Arctic Council into the future to cope with the Arctic issues. These state 'wants' are superior to everything else.
>
> (Negotiator 7)

This negotiator does not only speak about other actors' preferences for cooperation, as he has experienced it, but signals such wants to reflect the sole basis for the AC interaction. Another delegate explains it further:

> If you look back at the creation of the Arctic Council throughout today, we can see the level of cooperation is rising. Political cooperation. If we look at other regions, our countries have gone through better and worse, but the Arctic is always a peaceful, secure region with no conflicts. [...] I think this is pushing Arctic relations, the understanding of us having something that is peaceful and secure.
>
> (Negotiator 11)

Through such an understanding, we will arrive at a structure of expectations that not only involves a firm commitment to cooperation, but also to a cooperation that contains certain values to be cherished. It should be 'good cooperation,' and it should take place amongst peaceful members in a secure setting. It calls for trust among members, where appropriate behavior equalizes a behavior that does not violate peaceful and secure relations, and where everyone can count on others to follow. By dedicating one speech after another on commonness, trust, and shared interests and challenges, the Arctic states reinforce a structure where cooperation is prescribed. TFOPP delegates describe a process where all states have a say, and where everyone is making an effort to give and take. This would also be a general AC characteristic:

> My general feeling working on the Arctic Council is that everybody is solution minded, trying to find solutions rather than creating confrontations. And working in that way you kind of get mutual respect for other views [...] If you have sound proposals, are well-argued and with documented information to back it up, then you get results. But you also have to be able to respect others' views and not throw them away even though they do not fit your own view.
>
> (Adviser 5)

Just like George Herbert Mead's play- and game analogy (1925; 1934), the Arctic states are engaged in a game of Arctic appropriateness. Here states learn to draw a connection between appropriateness and multilateralism. To *not* cooperate, i.e., to deviate, is not an option within an institutional setting constructed around finding solutions. Instead, states stay put. Perhaps not so much because they have learned something new normatively on, for instance, oil pollution prevention, but since each Arctic state reflects upon its Self as an effect of social interaction. For commercial purposes, it is strategically important to stay in cooperation, and thus to act appropriate on that account. But being a trustworthy cooperative AC partner is also held important, which means that states cannot deviate from cooperation without causing a conflict in role conceptions: if not an Arctic cooperator, then a state's role – in relation to the Arctic – would fall apart. Really, in all of their Arctic strategies, the Arctic states are emphasizing them as 'being' Arctic, while drawing causal links between a well-being of the region and ensured cooperation. Therefore, to preserve the Arctic and keep it relatively healthy is a normative value also reflected in Arctic role-playing. However, to be oriented toward a common and general goal of cooperating, and toward a specific goal of oil spill prevention, does not mean there is a shared attitude regarding *what substance* to give the cooperation. In the next section, some dimensions of arguing processes in the TFOPP will be described.

### When Sovereignty Embarks – Arguing and Disrupted Commonness

The mandate upon which the TFOPP was founded was not a very concrete one; the Ministers handed over an assignment for the negotiators to "develop an Arctic Council Action Plan or other arrangement on oil pollution prevention" (*Kiruna Declaration*, 2013). Delegates then faced the task of deciding what type of document they were going to negotiate, and what they should include in the words and practice of prevention. Doing

that, actors turned to arguing. One delegate describes the negotiating process in the following metaphorical way:

> You know, you may have heard of the term "forming"? You form
> a group, and then it is "storming" after the group has been formed
> and people meet each other and its opinions, individual strategies, or
> stakeholder's perspective that are coming out and you storm through
> the process: I might have a different opinion from you and somebody
> else might have a different opinion from both of us, and we are going
> to disagree initially but by disagreeing, we eventually come to an
> agreement [norming]. And…is it always smooth? No, it is not always
> smooth. Is it what to be expected? Absolutely!
>
> (Negotiator 9)

As it would turn out, states argued, in particular, on three things: (1) how
far 'prevention' was about to stretch; (2) whether any legal obligations
should be attached to the document; and (3) the significant other, which in
this specific context would be represented by Russia.

*Arguing on Prevention*

Although there are Arctic stakeholders that argue prevention would
equalize a ban on Arctic oil production, most notably Greenpeace through
its still ongoing *Save the Arctic* campaign launched in 2012, projecting an
oil ban was not how prevention was approached in the TFOPP. Prevention,
according to the TFOPP, is a means rather than a goal, preventing *incidents*
and not oil-related activities per se. It does not mean to close down potential risk activities, but to operate within current legal frameworks, knowledge structures, and technological opportunities, *to lower* the probability
of an oil spill. The TFOPP result, i.e., the framework plan for cooperation
on prevention of oil pollution (2015), states: "recognizing that the prevention of incidents leading to the release of oil into the Arctic marine
environment is one of the most effective measures to protect the Arctic
marine environment." This TFOPP perspective gets support from both
the industry and the WWF, where the latter often provide information to
governments on Arctic environmental issues.

WWF calls for an avoidance of oil-related activities in especially sensitive areas. Furthermore, they desiderate technology and clean-up methods
that can handle oil accidents *when* they occur – because accidents *will* occur,
for which the industry currently is not considered fully equipped (WWF,
2017). Current technology and equipment does not allow for a successful
clean-up in cold temperatures and icy conditions, and infrastructure would
not be sufficient – for instance, would it be practically impossible to find
the amount of personnel and vessels needed (potentially with ice-breaking
capacities), that also are capable of operating in the icy water (EPPR, 2013;
interview 141208; 141210c; 150521a). The O&G, on the contrary, disagrees

and claims such technology to already exist, and considers itself to invest huge amounts in continuously developing new ones.[9]

In TFOPP, states were positioned along the above spectrum. A delegate with long experience working with oil and gas regulations, also within the industry, highlights the interpretative aspects of prevention:

> You will get back to the question of how safe is safe. The guy who is paying the bill will have one level of what he thinks is safe enough, whereas people who are going to be protected will have another level of what is safe enough. It is never going to be safe enough for the organizations who, you know, have an interest in just preserving the environment and it is always going to be too expensive for the regulator community. But our job is to walk the middle line and you know; the truth is almost always between those opposing points.
>
> (Negotiator 3)

Already at the first meeting, task force negotiators agreed that prevention, i.e., lowered risk of incidents, should be dealt with through cooperation, identification of challenges in Arctic operations, and bridging gaps in measures (TFOPP Two-pager, Oslo, 2014). Inputs were given by both WGs and the industrial community through presentations from, for instance, Gazprom, Rosneft, and Statoil/Equinor. For the next meeting, delegates decided to work with prevention within marine petroleum activities, and marine traffic (TFOPP Two-pager, Reykjavik, 2014; SAO meeting Yellowknife, 2014a). One of the co-chairs, the Norwegian Ambassador (SAO) Else-Berit Eikeland, elaborated on prevention as regulated differently depending on the sector: maritime transport would be internationally regulated (through IMO and its different conventions), whereas petroleum would be nationally regulated, more closed for insight. Eikeland thus suggested:

> Perhaps, when it comes to petroleum, the Arctic states have an interest in learning about our separate efforts, and then learning how we can exchange information and data, arrange meetings between our regulators, and examine safety culture…all the different elements in the prevention strategy.
>
> E-B Eikeland, SAO, Norway, co-chair TFOPP, 2014[10]

In TFOPP, preventive measures were discussed in relation to two areas: petroleum-related activities and maritime activities. Most of the measures that later was included in the Framework plan pertain to different kinds of information sharing across borders; indeed, to have industry standards harmonized over national borders was a key objective of Arctic cooperation on oil spill prevention, and of the TFOPP (150318b; 150521b; 151208). In relation to maritime activities, sixteen preventive measures were agreed upon that make out an Arctic-specific complement to the already established governance structure of foremost IMO and the Polar Code: whereas the Polar Code focused on vessel construction and how vessels are manned and equipped, the TFOPP objective was to add traffic management and to consider how vessels also could interact with each other to enhance safety precautions to prevent oil spills (interview 150709). In relation to the newer and nationally governed field of petroleum, the Arctic states agreed on five preventive measures.[11]

The preventive measures identified by TFOPP do, overall, encourage states to communicate. However, the final agreement also notes that information sharing should be in accordance with the respective state's national legislation, thereby giving states the right to withhold information should it be unreasonably difficult or costly to provide (Framework plan, paragraphs, 1.3–1.4). Prevention, as defined by the TFOPP, is therefore something that occurs between fully sovereign nation states. "We were not asking people to change how they conducted business," a delegate recalled, "it was just how to make it safer – nobody wanted to say to another state 'look, you have to do it better'" (interview 150626). Consequently, certainly, oil pollution prevention is about environmental protection, but it also has a strong dimension of state willingness included: how much are states prepared to reveal, exchange, and learn from each other for the environmental protection cause? According to one delegate, such willingness did not stretch very far: "[it became] absolutely clear that [states] were not ready to include anything in the agreement that went outside the national legislation or conventions that they already had signed up for […], sad, of course" (interview 151207). A pragmatic account of what to consider as a viable achievement is provided for in the following:

> Prevention… What it is really about is best practices and safe operating guidelines, and it is how the industry looks at health and that sort of safety culture. So, it is not so much what the governments are doing between themselves as how governments are regulating their industries within their national borders. We realized we were looking for a framework that would develop these best practices and that potentially would get national governments talking to each other about what they were doing, sharing information.
>
> (Date 151221)

Listening to this delegate, whose reasoning also gets support from the Framework agreement, and the arguments carried out by the TFOPP participants, we can conclude the oil spill prevention discourse to be about sharing information through enhanced communication channels, and not about putting restrictions on the oil and gas sector. Through such an understanding on prevention, the Arctic states simultaneously managed to avoid having to deal with the topic of oil-related activities per se, a discussion that would turn cooperation more – and perhaps too – political. Because, based on the oil spill prevention discourse, a conclusion could be drawn that with regard to Arctic fossil extraction, the Arctic states relate a bit too differently to the social world (and the value of oil extraction) to manage to point out a common route on preventing oil spills by going at the source.

*Arguing on the Legal Status of the Agreed Document*

At the first task force meeting, both Norway and Russia took on a leadership role and presented a draft on prevention measures and a suggestion for agreement, respectively (TFOPP Two-pager, Oslo, 2014).[12] Whereas Norway's draft on preventive measures was an area where arguing, meeting after meeting, led toward greater agreement and finally consensus, Russia's draft 'framework for agreement' was trickier. Because, what type of document were they set to negotiate – a framework plan, a memorandum of understanding, or perhaps an instrument to be legally binding? It would take until the fourth meeting before delegates had agreed that the preferred legal status of the document would be non-legal (TFOPP Two-pager, Helsinki, 2014). Delegates gave witness of a process suffering from a lack of clear direction. As one delegate explains, "the legal status of the document is part of a superordinate discussion that also governs content...once the final decision on title was made, discussion became more goal-oriented" (interview 150510). Another delegate explains the process as becoming more and more drained on ambitions as a direction could not be settled:

> To reach a consensual agreement, we realized we needed to adapt and make adjustments that would also lower our ambitions. That is, if we had emphasized our national position the whole way through, it would have been very difficult to come to conclusions. Gradually, it became more important to actually get an agreement, then what de facto was to be in it.
>
> (Adviser 3)

Whereas expectations initially had been high on what to achieve in the task force, the insecurities brought about by the negotiation process, of not knowing where it was headed, consequently led to lowered ambitions. Many participants gave witness of others wishing for more. Still, given all the storming that surrounded the legal status of the document, the result was quite good, they also continued, *if* just a proper follow-up on implementation would follow. Actually, as it turned out, states did initially not really have a fixed position on this issue, but were instead keen on following a consensus, whatever a consensus would imply.

For instance, Iceland, understood by others as initially searching for a legally binding agreement, explains its position to be the result of open-mindedness where it did not want to decide on the matter before knowing what would be included in the document. Similarly, Finland, who firstly spoke in favor of a binding agreement took on a different approach as it became clear what the majority preferred: "[i]n the beginning, it was not that much of a difference. But, when we first noticed there would not be a binding document, we also referred to not having the mandate to negotiate a legally binding agreement" (interview 150521a). Norway was perhaps the state that was most ambivalent to the issue – or – as the delegation explained it – lacked any strong preference in either direction but wanted to follow a consensus: "in that discussion, we kept in the background. Those who had strong emotions, strong wishes from their home country, those were instead controlling the discussion" (interview 150318a). The only thing that was for sure was that Russia was understood by all delegations as the state with the clearest ambition of signing a legally binding agreement at the end of the TFOPP process.

What is interesting here is not to decide which state held which position, but that most states gave an account of being willing to listen to the 'general view,' to take in arguments on the issue, and to see what others – in this particular case – preferred. It was expected to turn to discourses in order to establish a consensus on the issue:

> The night before the second meeting, we had a meeting together with other delegations. It was Denmark, Finland, Sweden, and the U.S. And we sort of mapped out where do we have common interests and common views? The core finding was that we actually see things similarly in most cases. [...] We wanted to map out the view of other countries regarding the legal status of the document.
>
> (Negotiator 4)

There might be several different reasons as to why these states preferred a non-binding agreement; a practical reason of simply lacking the mandate to negotiate judicial matters would be one; not having sufficiently strong interests in the negotiated topic could be another. However, a third reason seems to be that by not attaching any legal obligations to the document, it would be possible to go further in ambitions. A common thread in the interviewees' reasoning around a non-binding agreement was normatively motivated by a belief of it going to *benefit the cause* of oil prevention: if the agreement should carry a legal status as binding, the eight Arctic states would be capable of agreeing only on a minimum level for prevention. If the legal status instead would be loosened up, it would be possible to be more proactive, and where perhaps also the industry would be more willing to engage (interview 150506; 151127; 151221). Yet, there is one more plausible reason – not provided for by delegates themselves but argued next: Arctic states did not want what Russia wanted.

### Arguing due to the Significant Other

In Chapter 2, it was discussed how Habermas (1984) approached understanding as divided into three separate parts, where understanding and knowledge relates to the world in an objective, social, and subjective sense. The chances of arriving at a stage where actors are willing to take firm action increases the more common understanding there is between actors and in relation to the different worlds. So far it could be argued that 'staying put' in cooperation as well as perceptions of oil spill prevention as being an important issue, both express a shared common understanding in relation to the objective world. Less commonness is found in relation to knowledge and understanding in the social world, where interpretations of what prevention should mean and *how* it should be achieved divide the Arctic states. With Russia understood to be 'the significant other' a subjective world was also revealed, where lack of trust impacted upon cooperative ambitions.

In the TFOPP, Russia did not represent the same values as the rest. Therefore, the others distanced themselves – foremost by not backing up Russia in its key ambition regarding reaching a legally binding agreement on oil spill prevention. The identification against Russia should be understood in the context of timing. At the first TFOPP meeting, delegations were given the homework for the next meeting, i.e., to have clarified their national positions on which legal status should be attached to the document (TFOPP Two-pager, Oslo, 2014). Then, Crimea was illegally annexed into the Russian Federation. The violent acts in Ukraine were internationally condemned and gave rise to strong reactions. In time for the second TFOPP meeting, which took place only two days after the annexation of Crimea, the EU and many Western countries implemented strong sanctions against Russia (European Council, 2018). Although negotiations in the TFOPP continued as 'business as usual,' TFOPP was not unaffected. At least the Russian delegation itself linked Sweden's, the U.S.,' and Canada's negative attitudes to a binding agreement to what had just happened in Ukraine. The political situation was interpreted as the reason.

Within the interactive structure in the Arctic Council, there is a strong support for peace and security. To violate this, especially if it takes place outside of the Arctic Council per se, is to gravely violate what is considered appropriate, as in legitimate and justified. The alter part of the roles played by the Arctic states, therefore, quickly became aware of the prevalent expectations on taking a distance from such behavior. Thus, rather than preferences on the specific issue per se, social relations helped to guide the Arctic states to their 'wants' in the TFOPP. Russia, who already views itself as somewhat different from the rest, as someone who – although dependent on cooperation – follows its own route toward future Arctic leadership, can cope with being the significant other without causing too much of a disturbance in its role.

To go back to Habermas and understanding in the subjective world: Russia's strong emphasis on a binding agreement and reluctance to let this ambition go, together with its Arctic policy and interests, made other states question the sincerity of Russia, not knowing what to expect. One could assume the Arctic states to reflect on Russian motives for engaging in oil pollution prevention, and whether words expressed could be trusted. This had not only to do with Ukraine, but also with a poor environmental track record with as much as several thousands of small-scale leaks annually, where about 500,000 tons of these oil spills are calculated to annually end up in the Arctic Ocean, where technology and infrastructure are insufficient, and where leaking pipes remain in the ground for too long (Vasileyeva, 2011; Greenpeace, 2017). Furthermore, it could not be ruled out that other political state interests were hiding underneath oil pollution prevention, similar to what has been suggested to govern Russia's earlier active participation in negotiations on the Polar Code (Bognar, 2016).[13] One delegate illustrates a careful attitude toward Russia in the following way:

> There are countries with higher tolerance levels [on dangerous spills], like Russia, even though they, of course, would never admit it. It then becomes difficult when trying to implement preventive measures that require change, inference with current regulations. Above all, [prevention] is very much about access to information, like exchange of satellite information, weather forecast, and similar. And, if then a suspicious…if one is sort of suspicious regarding what the information will be used for, then obviously one will be very careful regarding what to agree to. What is written on paper? Because one might think the information will be used in some other way, for instance.
>
> (Adviser, 2)

Although the Arctic state delegations appeal to reason and professionalism as a way to distance Arctic cooperation from external happenings, political conflicts were lurking behind the surface, making Arctic states cautious: "[…] at the end of the TFOPP, we thought we could notice the tension, even though it was not that big, but one could notice a slight deterioration with regard to the atmosphere. More suspicious than in the beginning" (interview 150521a). Despite this, social relations were united in a common understanding on 'staying put,' which fought back on political turmoil and wanted to keep the focus on oil spill prevention. As such, it was a negotiation process that state representatives above all wanted to describe as good and constructive.

## Continuity and Change in Role Performances

The state performances in the TFOPP followed the role pattern discussed in Chapter 4. It was the coastal states that had front positions. Above all, it was Norway and Russia who acted as leaders: Norway as the know-how leader that wanted to spread its competencies and led the way, and Russia who in TFOPP found a way to inform on how cooperation most suitably could be organized paying respect to sovereignty. The U.S. also acted like a leader, although to a lesser issue-specific extent and instead more 'innate' to its character, assuring everyone as being equals and by cherishing best practices for everyone to learn from, and Canada followed its role of being a protector. The other four states were less visible: Denmark took on a pragmatic (coastal state) kid-brother role and did not reveal any clear goals of its own or attracted attention; Iceland was the follower with ambitions, who wanted to listen to a consensus but felt pleased over its ideas being well accepted; Sweden and Finland took on low-key positions within an area they felt was not for them to direct.

That continuity rather than change marked states' role performances should not lead to the conclusion that states were not subjected to learning: a modest result regarding the environmentally protective strength of the agreement does not exclude the occurrence of social learning. Indeed, the argument underpinning this book is that environmental protection is less determined by the persuasiveness within the negotiated issue per se, and more by the social relations guiding the negotiations. In the remaining part of this chapter, three examples will be given where states have shown sensitivity to others' expectations and thus proven themselves – possibly – receptive to learning. Others' expectations do not matter here because these states lack a will of their own, but because they need these others' expectations to make sense of their role.

### Confirmative Role Behavior by Finland and Sweden

Had it not been for knowing about Finland and Sweden's role performance in the AC, one could have expected these two states to push for strong oil spill prevention measures, due to both being environmental norm entrepreneurs not directly involved in the Arctic oil and gas extraction. Such behavior would, however, go against the prescribed AC behavior, where sovereignty is to be respected. In relation to discussions on petroleum measures, a negotiator explains their position:

> We made a conscious decision that when the Task force discussed the petroleum part, then we were mostly to follow discussions. [Sweden/ Finland] did the same, which is very wise because we are not oil producers so we really should not go there and say "well, we think you should do this or that." [...] You know, there is a term called "besserwisser," from German. Why should one go and comment and criticize on anything one does not really have responsibility for in real life?
>
> (Date 150603)

Both Sweden and Finland played the roles of supporting others; they spoke when they had knowledge to contribute and remained silent in between. They were team players, respecting the sovereignty of others while being ready to contribute and teach insofar as there was a request thereof. As another delegate expressed, there is a general expectation within the AC to shoulder responsibility within areas that one knows about: "I believe expectations are the same for all members, namely, to contribute with those things one knows of" (interview 150510). This was a prescription both Sweden and Finland followed. By the same token, they received positive recognition:

> We had strong opinions within those issue areas where we knew we were skillful. Maybe equally skillful as the big nations, least. Like icebreakers, for example. We are in the front edge there. We cooperate on ice management issues and act basically as one big organization during wintertime, and environmental surveillance by air where we are world leading, for instance. Those are the things we have brought up, instantly having received positive feedback: "this is good, this is something we can work into our common document." It is there we have focused; on things we know we are good at.
>
> (Date 141218)

Another delegate continues by describing what would be noticed as expectations on how to act: "[w]e had the feeling that whenever there was a need for something constructive or for a peacemaker, we were asked for a comment or view" (interview 150603). Interaction thus confirmed these states to be what they themselves expected and conceived to be, indicating no need for a role change. Rather, their roles were firmly rooted in a social structure where others' recognition boosted their importance as actors, if playing their roles correctly.

### A Canadian Attempt for Alter-Casting?

In the Arctic Council, Canada performs as the 'Protector.' Generally, this role concerns the indigenous population, but it also moves beyond that to a protection of important principles. In the TFOPP, injustice came knocking on the door through the Ukraine crisis, calling for the need to protect international law. Canada activated a role performance on standing firm against grave violations of international law.

Outside of the TFOPP, but yet within the AC, Canada together with the U.S. and North American–based indigenous organizations boycotted a task force meeting on the reduction of black carbon and methane, hosted by Russia. Canada also chose to cancel a showcase event planned to take place in Ottawa in April 2015, just a day prior to the big Ministerial meeting in Iqaluit marking the end of the Canadian chairmanship period. The showcase event had previously been prioritized by Canada as an outreach activity. WGs were supposed to present their work, and cancellation caused surprise and irritation (interview 150510). In media, the cancellation was reported as having to do with the continuing tensions over Ukraine, and a worry over the symbolism of having senior Russian officials wandering the streets of Ottawa (McDiarmid,

2015). A person with insight into the showcase event confirmed: "the last government [Harper government] was really sensitive about Russia, and did not want to have you know, media coverage of Canadians and Russians together, *in Ottawa*" (interview 151127). Also, Russia sent an answer back in similar style: less than two weeks prior to the Ministerial meeting, it was announced that Sergey Lavrov – attendee at the Arctic Council Ministerial meetings since 2004 – was unable to come due to "prior commitments" (Arctic Yearbook, 2015). In both cases, the meeting scene was politically orchestrated, spectacles in the sense of not wishing to – presenting imageries of – friendly partaking.

The diplomatic turmoil could also be noticed within the TFOPP. While strengthening its own role as a protector, Canada attempted to alter-cast Russia's role as un-trustworthy. Indication of this was expressed in two different ways. Firstly, a common thread found in the interviews is the informal, more personal, aspects of negotiations, where, if not trust, at least some sense of understanding was developed: to have dinner or a beer is understood as a good setting for discussions to thrive and understandings to increase. However, to engage in such activities with Russia was not seen as appropriate by Canada, who did not want relations to appeal to "commonness" but rather differences. A delegate explains:

> The Canadian negotiators had clear restrictions regarding how to behave when leaving Ottawa. They were, for instance, told not to associate with Russian representatives, even on an informal basis. And that is quite unusual. And then, we had a negotiation in Ottawa a year ago, when it took a very long time before the Russian representatives were permitted a Canadian visa. I think these were things done deliberately, to provoke.
>
> (Negotiator 7)

The second way to signal Russia as being different from the rest was to simply not agree with the state, signaling it as coming from a different 'world.' For instance, since a legal agreement turned out impossible, Russia held on strongly to its last bedding straw, of having Ministers at least sign the document, although it would not be binding. However, Canada was described to refuse any such signing. Here, the importance of story-telling and 'optics' should again not be underestimated; a signed document would send signals of friendship, which Canada at that moment did not want to emphasize or orchestrate.

For Russia, the situation was reverse. Here story-telling on trust and partnership – through signatures – would indeed have been valuable. For similar reason, Russia also avoided to take on the offender role suggested by Canada, and kept promoting values of cooperation, mutual trust, and unique relations. As it seems, this role found support all the way to the Ministerial level. When the Canadian chairman declared to bring Ukraine up at the Ministerial meeting, the U.S. Secretary of State John Kerry had a different approach: although concerned over the security situation, he referred to a recent phone call with Foreign Minister Lavrov:

> [...] he made it crystal clear to me that Russia wants the council to be successful, that they want this to be a cooperative entity that is geared towards peaceful purposes, and that it's their intent to cooperate with us on the protection of the environment on the agenda we have set forth [...].
>
> (Kerry, 2015)

The trust offered to Russia does not only include the state within Arctic commonness but raise the cooperative stakes: Russia was faced with expectations of what would be considered an appropriate behavior: peaceful and cooperative manner. The fact that Russia chooses to play this role – and not deviate when they met resistance – indicates a socialization process at play in the Arctic Council, continuously strengthening cooperation as appropriateness, no matter which (material) interests one has of the Arctic.

In the TFOPP also, Russia cooperated and its (political) impact was described as less could be expected. In addition, and somewhat surprising, Russia also brought an environmental language to the draft, pushing for ecological risk analysis (interview 150318b; 150521a; 151127). Nonetheless, it could be argued that Russia responded *somewhat* to those offender expectations that encircled interaction: delegates give an account of a Russia that became more self-confident than before, whose delegation started to act in a disturbing manner with little consideration for others, who jumped back and forth between paragraphs, and proposed silly changes to minor things. The irrational behavior, as it was understood to be, made delegates start to question the purpose, even reflecting on whether this small micro-behavior would be a power demonstration of being 'back on the international scene' (interview 150603; 160623a).

Russia considers itself different from the West and the rest and foresees itself as becoming an Arctic leader already in the years 2016–2020, making full use of its strategic resource base (Arctic strategy, 2008:11a–c). To phase setbacks in strategically important issue areas could, for a leader with hyperborean origin, very well activate a behavior that is perceived as annoying and self-assertive by others, no matter how important cooperation – as a means – yet may be. With an interactive structure allowing for differences rather than commonness, appropriate behavior would not necessarily any longer be associated with the value of Arctic cooperation. Instead, following Mead (1925; 1934) the cooperative disturbing behavior could be a response (by *I*) to those alter expectations that also expected such a behavior.

## Norwegian Role Incoherence

Norway is an Arctic state characterized by being both environmentally proactive and a major oil and gas producer. In the TFOPP, the Norwegian state delegation's representation came from ministries such as transport and communication, labor, foreign affairs, trade, industry, fisheries, as well as from the petroleum safety authority. This is telling of a Norway that within the TFOPP context aimed to play the role of a knowledgeable oil producer and maritime actor. Yet, due to others' expectations, Norway came to suffer – for a short while – of a role incoherence, allowing for a short guest performance by its environmental *Me*.

The concrete situation concerned ecological mapping: was this a good preventive activity to be included within the otherwise industrial-focused measures listed? Norway was described to be clearly opposed to such a measure and explained oil pollution prevention to not be about environmental protection activities, but about measures related to regulators, industries, and shippers' ordinary activities. It was a position that caused other delegations to react: "First, they did not want to have anything written about the vulnerability of the Arctic environment. They stated it was not the idea of the whole paper to talk about how vulnerable the nature is" (HoD 4). Norway's reluctance to bring the environmental aspect caused disagreement, where others foresaw that *not* drilling close to environmentally sensitive areas indeed would be a prevention measure (150521a; 151127). Then, however, Norway suddenly changed its mind:

> They [Norway] said, "it will never be in the document, never ever, over my dead body." But for the next meeting, they had changed their views and were collaborating on this. They even had done a lot of background work to phrase sentences on this topic in such a way that they could accept it. I don't know what had happened there in the background, between meetings, but obviously something.
>
> (Negotiator 4)

What is interesting here is, foremost, two things. Firstly, the initial disappointment on Norway's reluctance to incorporate writings on ecological mapping is telling of what expectations others had on the country: to *not* explicitly promote the environment was not among them. Secondly, when confronted with these expectations of being someone who *should* be acting in the name of the environment, Norway did change its position. The Norwegian delegation explains the change to be a concession of being in a negotiation setting where a consensus would be the guiding principle:

> I think some may have perceived us as being a bit stiff, in that specific issue [ecological mapping and environmental risk analysis]. But, it was simply those formulations that were on the table, they would be

problematic for us to handle within our system, a system that we argue is very good at providing for environmental considerations. [...] In the beginning, we therefore were against it, it breached the structure of our document – it did not belong there. But, since it was important to some and since we had to reach a consensus...in dialogue with our Senior Arctic Official, we therefore chose to stand back on that one as well.

(Adviser 3)

To Norway, it was a rational sacrifice, logical to do in order to bring the agreement ashore and keep relations smooth. However, there might be more to this change in position than simply strategic reasoning. Being one of the world's top suppliers and producers of oil, Norway is an important player at the global energy market; furthermore, it has much to thank the oil sector for its national welfare (Jensen, 2016). Whereas the Norwegian oil production in general was declining between 2001 and 2013, the years since 2014 have witnessed a steady increase. Energy extraction and activity in the coming fifty years is also expected to be high, with close to 50 percent of all resources still expected to be recoverable. Of these undiscovered resources, more than half (64 percent) is located in the Barents Sea (Norwegian Petroleum, 2020). Yet, Norway may be as important a petroleum nation as it may wish, but delegations within the TFOPP expected it to include more things than a self-interested attitude toward oil and gas, such as a promotion of environmental values. By not following that, Norway gave clues for odd behavior, illustrated by the following quote where a delegate gives voice to the surprising factor within the TFOPP:

...It is *always* Norway. Norway is framing the discourse that "it is not the Arctic, we are not really Arctic, we can drill." And, it is always very surprising to me how the Norwegian government is very clear about "we can drill and the measures are in place and we will sponsor piles of report that will show it [...]."[14]

(Adviser 9)

If Norway had just been about petroleum, it should have come as no surprise that such a state wanted to drill. However, there were other expectations as well, stemming from the domestic sphere. To compare Norway and the U.S., for instance, far few Norwegians support Arctic fossil extraction than their counterparts in Alaska (Langhelle and Hansen, 2008). Similarly, to explore the Barents Sea might be, as described by Equnior, a "safe and responsible thing to do" (Equinor, 2019), and yet there is "skepticism and opposition" toward such activities (ibid.). The decision by the Norwegian government to open up the Barents Sea was also met by political criticism, as well as a lawsuit from parts of the environmental movement (Greenpeace, 2016a; Finne, 2015). There is, thus, a Norwegian role incoherence.

Norway's sudden change in the TFOPP could therefore be a result of colliding *Me*s. If *Me* is the part of the Self that views itself through the eyes of others (Mead, 1925), then the situation became problematic since it challenged core understandings on what type of actor Norway would be. Whereas it was Norway as a petroleum nation (and Arctic maritime actor, in general) that acted in the TFOPP, it met expectations that targeted Norway as a leading normative power within the environmental field. Although Norway was not playing this role in the TFOPP (it was perceived the wrong forum, environment was not what they discussed here), one could not walk freely from these alter expectations. One could not walk free from them since they were also part of the Norwegian Self. When Norway viewed itself through the eyes of others, it became aware of expectations linked to the "green" version of its social *Me*. Although the forum was perceived to be wrong, the impulsive and inner *I* quickly stepped in to make sense of *Me* and reacted as had it been the normative 'doing good state' who responded. It thus corrected the role behavior by taking on a slightly different – and more flexible (i.e., appropriate in relation to expectations) role.

## Conclusion: Learning in Negotiations on Marine Oil Spill Prevention

The discussion within this chapter has centered on oil pollution prevention – its cooperative background and how the TFOPP process reached a consensus on the matter. Theoretically, it has applied a processual approach where a macro-level object such as that of an international negotiation on an issue of environmental importance has been investigated utilizing a micro-level perspective. As such, the chapter has shed light on how and to what degree the Arctic states have come to view national action-taking on marine oil spill prevention as a valid course of action. Furthermore, it has offered insight into the social process where states – through their assigned representatives – come to reflect and pay attention to its role, which includes others' expectations on how it should be performed.

The negotiation process on Arctic marine oil pollution prevention was proved to be a process controlled by different interactive logics and interests, depending on which dimension of understanding – objective, social, or subjective – that was in the analytical spotlight. In this chapter, it has been argued that a common understanding marked state interaction regarding topical importance; to cooperate and find an agreement on oil spill prevention was an objective goal, i.e., a *value*, shared by all members. More difficult was then the issue of which measures should take them toward this goal, and how. Settling this, actors turned to arguing to define the prevention discourse, and how it should be approached legally. Arguing did not lead the actors cooperatively astray, but it unfolded different understandings in relation to actors' social and subjective worlds; socially, experiences were not shared to the degree that just one appropriate approach to oil extraction and environmental protection was viewed as valid, and subjectively, trust also was failing.

From a longitudinal perspective, the fact that the AC has arrived at cooperation on oil pollution prevention is indicative of a progression on environmental norms to have occurred in international relations, which furthermore entails a learning on the human-nature nexus. However, the timeframe within TFOPP was narrow, consequently with modest or no support for the occurrence of a learning on the norm of oil pollution prevention. Still, although TFOPP did not affect states in such a way that they changed their mind concerning the value of norm, the process yet entailed learning; states learned of their social position – their role – in the AC cooperation. By the same means, states received information about how to relate and position themselves to the issue of oil pollution prevention.

This chapter has dealt with processes of consensus-reaching and role-playing as both being interactively informed, sprung social relations. Expectations – ego, alter, and others – and recognition have proven to be important role-behavioral components by appealing to actors' senses of Self. In the chapter, this has partly been illustrated using three examples of states' role performances as decided (somewhat) in accordance with others' expectations and with prevalent prescriptions of a cooperative behavior. In the first case, drawing on the role performance of Finland and Sweden, alter and ego expectations coalesced, causing little friction in roles. In the second case, also, illustrated through Canada-Russia relations, alter and ego expectations coalesced. However, this time, social interaction revealed (a view of) Arctic relations as being different and unilateral rather than held in common, thereby extending the inter-active structure to motivate and allowing for a rather 'tough' protector role, as in line with Canada's role conception. In the third case, drawing on Norway's role performance, others' expectations led to a changed (adapted) position in relation to a specific issue. This time social inter-action revealed an internal role conflict, whereby Norway could not have its performance confirmed by alter. To reach a role coherence, Norway was pushed toward others' expectations, flexing its role to be under-stood – by others as well as the Self – correctly.

In all three examples, states turned to alter to have 'the accuracy' of their performance confirmed. Thus, by considering alter expectations, actors highlighted those aspects of themselves that were perceived to fit the current interactive structure the best. Although this would not auto-matically trigger a change in role behavior, it presupposes the existence of state reflections which nonetheless is being indicative of a learning process at play. Thus, as the title of this chapter suggests, although sovereignty was expected to interfere and restrain the negotiation process between the Arctic states on marine oil spill prevention, a micro-level analysis reveals nuances by identifying state interaction to include also normative prescriptions and social receptiveness.

# Notes

1  The material for this chapter mainly draws on interviews. For this book, just under 30 formal interviews have been conducted, in addition to half a dozen informal interviews, with state representatives and other actors with experience of AC cooperation. Of the formal interviews, 16 have been conducted with TFOPP participants specifically, of which the majority were state representatives. Each state delegation consists of a HoD (from the relevant government ministry) acting as chief negotiator, in addition to civil servants and experts from relevant ministries and authorities. In TFOPP, the business sector was invited to join delegations as well, for instance, oil companies. Except from Canada, where none of the contacted state officials wanted to be interviewed, the remaining seven states are represented.

2  Norway discovered the offshore Snøhvit gas field in 1984. The giant Russian gas field Shtokman, as well located in the Barents Sea, was discovered in 1988, and the Prirazlomnoye oil field in the Pechora Sea was discovered in 1989.

3  For instance, the symbolical link between polar bears and climate change reportedly made George W. Bush Jr seriously consider a deregulation of the Endangered Species Act, fearing that a listing of polar bear as endangered would cause greater debate and more forceful demands on reduced greenhouse gas emissions (Goldenberg, 2009).

4  The perspective from the Arctic Athabaskan Council was a bit different, where oil and gas as well as other forms of resource extraction simply had left a mess (AC Ministerial meeting Barrow, 2000:1–2). In contrast to climate change, where all PPs have shared the view on urgent need for actions, oil spill prevention has been less unison. For instance, the Aleutian Island societies are a mixed economy: located along the heavy trafficked North Pacific great circle route, one fears the risk of oil spills. At the same time, one wants to have the economic opportunities of the region safely provided for, like fishing (interview 151221). Indigenous groups at other places are of different opinions. To save and protect the Arctic must not, from an indigenous perspective, be about decreased human activities. But it could.

5  Not just working groups but also the Arctic states – the SAOs – had brought the issue up on numerous occasions: Canada had informed on prevention strategies, Norway had conducted mapping on resources at risk from spills, and Russia had sent invitations and project proposals on marine accident prevention; all of them received positive feedback in return where cooperation has been understood as beneficial (SAO meeting Fairbanks, 2000; Narvik, 2007).

6  The Macondo blowout, when a BP-operated oil rig exploded and sank in the Gulf of Mexico, caused a three-month constant flow from the oil well, which would take four years to clean up with help from huge amounts of chemical dispersants. At its peak, close to 50,000 people and 6,500 vessels were assisting in the clean-up (BP, 2015). In the Arctic, conditions are different with few oil-eating micro-organisms compared to tempered waters, and where technology has a hard time functioning properly when the cold, i.e., ice, sets in. And, it is dark for many months out of the year.

7  Parallel in time, EPPR and PAME both presented thorough information on Arctic states, as a response to the oil spill in the Mexican Gulf: PAME *Safety Systems Management and Safety Culture Guidelines* (2014) and EPPR *Recommended Best Practices for Arctic Oil Spill Prevention* (2013).

8  This interview has been conducted by the Arctic Council and published on the Arctic Council webpage.

9  Information accessed from oil and gas conferences; Arctic Oil and Gas Conference, 2014, Arctic Technology Conference, 2015, and IEA Gas & Oil Technology, at Arctic Frontiers, 2017. In addition, a task force participant explains that the O&G industry also identifies weaknesses in its response capacity, and acknowledges the substantial risks and costs associated with oil spills. More proven technology therefore is needed that has the capacity to function properly also under Arctic conditions. Because, as this interviewee descriptively explains, "the Arctic and an ice-covered swimming pool are two completely different things, even if waves would be added" (interview 151207).

10  This interview has been conducted by the Arctic Council and published on the Arctic Council webpage.

11  Within the area of petroleum activity, safety measures and equipment are in focus, in addition to promoting and strengthening cooperation between the national authorities, i.e., "the regulator's community" (Framework plan, paragraph 2–2.3). The maritime area covers exchange of experiences, best practices, and lessons learned within areas such as: maritime traffic, the possible coordination of hydrography and mapping surveys, improved meteorological and oceanographic forecasts, improved satellite communication, ice-breaking services, and developing catalogues of existing resources (tug boats, ship arrestors, etc.). A preventive measure with a direct link to environmental protection is the measure to explore and pursue ways to reduce environmental risks posed by transportation, storage, or use of heavy fuel oil ships in the Arctic (ibid.: paragraphs 3.3).

12  At the second meeting, the U.S. also presented a draft document of a regulator's forum, aiming for exchange of best practices within the oil and gas industry (TFOPP Two-pager, second meeting in Reykjavik, 2014). The later became the AORF and is a concrete result of the TFOPP, although freestanding from the AC, dedicated to information sharing and best practices regarding offshore regulation.

13  Russia has an interest in upholding its jurisdictional claims and legal control over the Northern Sea Route, and the vessels sailing through, which is why its proposals for the *Polar Code* when under negotiation mainly served to – in various ways – ease its future as an Arctic state (Bognar, 2016).

14  The vast majority of Norway's oil and gas production has so far been extracted in the Norwegian Sea and the North Sea. The entry into Arctic drilling, through the Barents Sea, is thus a new activity.

## References

Arctic Environmental Protection Strategy (AEPS). (1991). *Declaration on the Protection of the Arctic Environment.* June 14, 1991, Rovaniemi, Finland.

Arctic Monitoring and Assessment Programme (AMAP) (2007). *Assessment 2007: Oil and Gas Activities in the Arctic.* Oslo, Norway.

Arctic Yearbook 2015. (2015). "Year in Review: 2014," in Heininen, L., Exner-Pirot, H. and Plouffe, J. (eds.). *Arctic Yearbook 2015. Arctic Governance and Governing.* Iceland: Northern Research Forum.

Bognar, D. (2016). "Russian Proposals on the Polar Code: Contributing to Common Rules or Furthering State Interests?" *Arctic Review on Law and Politics,* Vol. 7:2, pp. 111–135.

British Petroleum. (BP) (2015). *Gulf of Mexico. Environmental Recovery and Restoration.* Five Years Report: March 2015. Prepared by BP Exploration & Production.

Clark, W. C., Mitchell, R. B. and Cash, D. W. (2006). "Evaluating the Influence of Global Environmental Assessments," in Mitchell, R. B. et al. (eds.). *Global Environmental Assessments: Information and Influence.* Cambridge, MA: MIT Press.

Eikeland, E-B. (2014). Senior Arctic Official. *Interviews from the Task Force on Oil Pollution Prevention: Part 2.* Available via www.arctic-council.org/index.php/en/ component/ content/article?id=221:task-force-on-oil-pollution-prevention-interviews-part-2.

Emergency Prevention, Preparedness and Response (EPPR). (2013). *Recommended Best Practices for Arctic Oil Spill Prevention.* O. K. Bjerkemo (ed.). prepared by Det Norske Veritas (DNV) on behalf of the Norwegian Coastal Administration. Published by EPPR. Technical Report.

———— (1998). *Environmental Risk Analysis of Arctic Activities. Risk Analysis Report 2.* Emergency, Preparedness, Prevention and Response (EPPR).

Equinor. (2019). *Why It's Responsible to Explore the Barents Sea.* December 12, 2019. Accessed from www.statoil.com/en/what-we-do/responsible-drilling-in-the-barents-sea.html.

European Council. (2018). *Timeline – EU Restrictive Measures in Response to the Crisis in Ukraine.* Last reviewed January 8, 2018, accessed on January 13, 2018, from www.consilium.europa.eu/en/policies/sanctions/ukraine-crisis/history-ukraine-crisis/

Finne, A. F. (2015). "Regjeringen provoserer: Åpner omstridte leteområder og flytter iskanten." *High North News.* January 20, 2015.

*Framework Plan for Cooperation on Prevention of Oil Pollution from Petroleum and Maritime Activities in the Marine Areas of the Arctic.* (2015). Annex 3: Iqaluit 2015 SAO Report to Ministers. Tromsø, Norway: Arctic Council Secretariat.

Goldenberg, S. (2009). "The Worst of Times: Bush's Environmental Legacy Examined." *The Guardian,* January 16, 2009.

Greenpeace (2017). *Russian Oil Disaster. The Ongoing Arctic Oil Spill Crisis.* Accessed on June 12, 2017, from www.greenpeace.org/international/en/campaigns/climate-change/arctic-impacts/The-dangers-of-Arctic-oil/Black-ice – Russian-oil-spill-disaster/.

———— (2016). *Historic Lawsuit against Arctic Oil.* Press release October 18, 2016. Accessed from www.greenpeace.org/international/en/press/releases/2016/lawsuit-arctic-oil-norway-historic/.

Habermas, J. (1984). *The Theory of Communicative Action. Reason and the Rationalizaton of Society,* Vol. 1. Cambridge: Polity Press.

Hough, P. (2013). *International Politics of the Arctic: Coming in from the cold.* London and New York: Routledge.

Interviews. (2014–2015). Interviews conducted between December 2014 and December 2015 state representatives and other representatives active in Arctic Council cooperation and negotiation.

Jensen, L. C. (2016). *International Relations in the Arctic. Norway and the Struggle for Power in the New North.* London: I. B. Tauris.

Kerry. J. (2015). *Press Availability Following the Arctic Council Ministerial. Secretary of State John Kerry.* Iqaluit, Canada. April 24, 2015. Accessed via oaarchive.arctic-council.org.

*Kiruna Declaration.* (2013). Declaration of the Eight Ministerial Meeting of the Arctic Council, held in Kiruna, Sweden, May 15, 2013.

Kontorovich, A. E. (2015). "Oil and Gas of the Russian Arctic: History of Development in the 20th Century, Resources and Strategy for the 21st Century," *Planet Earth Science,* Vol. 41:2, pp. 42–61.

Langhelle, O. and Hansen, F. K. (2008). "Perceptions of Arctic Challenges: Alaska, Canada, Norway and Russia compared," in Mikkelsen, A. and Langhelle, O. (eds.). *Arctic Oil and Gas – Sustainability at Risk?* London and New York: Routledge.

McDiarmid, M. (2015). "Arctic Council Tensions Threaten Environment as Canada Exits Chair." CBC News – Canadian Broadcasting Corporation. April 25, 2015.

Mead, G. H. (1934/1972). *Mind, Self and Society. From the Standpoint of a Social Behaviourist.* Edited and with an introduction by C. W. Morris. Chicago: University of Chicago Press.

——— (1925). "The Genesis of Self and Social Control," *International Journal of Ethics,* Vol. 35:3, pp. 251–277.

MOSPA. (2013). *Agreement on Cooperation on Marine Oil Pollution Preparedness and Response in the Arctic (MOSPA).* Signed May 15, 2013, Tromsø, Norway: Arctic Council Secretariat.

National Ocean Service. (2014). *The Exxon Valdez, 25 Years Later.* Podcast. National Oceanic and Atmospheric Administration; U.S Department of Commerce.

National Oceanic and Atmospheric Administration (NOAA). (2017). *U.S. Department of Commerce. Gulf Oil Spill.* Last updated March 2013, accessed on December 30, 2017, from www.noaa.gov/resource-collections/gulf-oil-spill.

Norwegian Petroleum. (2020). *Exploration Activity.* Last updated February 2, 2020. Accessed from www.norskpetroleum.no/en/exploration/exploration-activity/.

Obama, B. (2010). *Address to the Nation.* Full text of President Obama's BP Oil Spill Speech. June 15, 2010, Washington. Accessed from www.reuters.com/article/us-oil-spill-obama-text/full-text-of-president-obamas-bp-oil-spill-speech-idUSTRE65F02C20100616Reuters.

Protection of the Arctic Marine Environment (PAME). (2014). *Arctic Offshore Oil and Gas Guidelines: Systems Safety Management and Safety Culture. Avoiding Major Disasters in Arctic Offshore Oil and Gas Operations.* PAME: Iceland.

——— (1997). *Arctic Offshore Oil and Gas Guidelines.* PAME: Iceland.

Rottem, S. V. (2016). "The Arctic Council in Arctic Governance: The Significance of the Oil Spill Agreement," in Jakobson, L. and Melwin, N. (eds.). *The New Arctic Governance.* SIPRI Research Report No. 25. Oxford University Press.

——— (2015). "A Note on the Arctic Council Agreements," *Ocean Development and International Law,* Vol. 46:1, pp. 50–59.

Russia's Arctic Strategy. (2008). *Russian Federation Policy for the Arctic to 2020.* Adopted by President D. Medvedev on September 18, 2008.

SAO Discussion Paper. (2016). *Discussion Paper on the Arctic Council's Work on Oil and Gas Issues.* Submitted by United States of America. Arctic Council SAO plenary meeting, Fairbanks, Alaska, March 16–17, 2016. Accessed from the Arctic Council Open Access Repository: https://oaarchive.arctic-council.org/handle/11374/1730.

SAO Meeting Minutes. (2018). *Arctic Council Senior Arctic Officials Meeting Haparanda, Sweden.* November 14–15, 2012. Final report, pp. 1–11. Accessed via oaarchive.arctic-council.org.

——— (2011). *Luleå, Sweden.* November 8–9, 2011. Final report, pp. 1–13.

TFOPP Two-pager. (2014). *Summary Two Pager.* Task Force on Oil Pollution Prevention. Meetings 1–5, January 2014–November 2014. Accessed from the Arctic Council Open Access Repository: https://oaarchive.arctic-council.org/handle/11374/75.

United Nations Conventions on the Law of the Sea (UNCLOS). (1982). UN General Assembly, December 10, 1982.

Vasiliev, A. (2014). Senior Arctic Official. *Interviews from the Task Force on Oil Pollution Prevention: Part 1.* Accessed from www.arctic-council.org/index.php/en/component   /content/article?id=219:interviews-from-the-task-force-on-oil-pollution-prevention-part-1.

Vasilyeva, N. (2011). *AP Enterprise: Russia Oil Spills Wreak Devastation.* Boston. com, December 17, 2011.

World Wildlife Fund (WWF). (2017). *Threats: Oil and Gas Development.* Accessed on August 18, 2017, from www.worldwildlife.org/threats/oil-and-gas-development.

——— (2014). *Modeling Oil Spills in the Beaufort, Bering and Barents Seas.* WWF Arctic factsheet.

# 6 When Action-Taking Is Prescribed: A Micro-Level Departure in Negotiations on Short-Lived Climate Pollutants

## Short-Lived Climate Pollutants in the Arctic

When assuming the Arctic Council chairmanship for the period 2015–2017, the U.S. Secretary of State John Kerry explained climate change to be one of the top three priorities for U.S. He explained how a decrease in sea ice causes flooding and makes coastal villages vulnerable to storms, how thawing permafrost is linked to wildfires and collapsing infrastructures, and how it may release methane, a gas twenty times more potent as a greenhouse gas than carbon dioxide. "The ability of future generations to be able to adapt and live and prosper in the Arctic in the way people have for thousands of years is tragically but actually in jeopardy," Kerry said, urging the Council members and observers to "do more on climate change" (Kerry, 2015).

Illustrative of the wish to 'do more,' the Ministerial meeting in Iqaluit in 2015 adopted an agreement titled 'Enhanced Black Carbon and Methane Reductions – An Arctic Council Framework for Action' (hereinafter the Framework Agreement). It had been negotiated by the Task Force for Action on Black Carbon and Methane (hereinafter the TFBCM), and it moved scientific knowledge and recommendations on emissions reduction into the domain of political action-taking. In principle, the agreement listed three different key commitments for the states – three 'to do's: (1) take enhanced, ambitious, national, and collective action to reduce emissions on black carbon, and significantly reduce methane emission; (2) adopt an aspirational quantitative collective goal on black carbon by 2017; and (3) carry out national inventories of the frequency and spread of these pollutants, and send biannual reports on existing and planned actions on emission reductions (Framework Agreement, 2015).

Shaped as a non-legal framework document, the agreement is not stringent regarding who should do what, when, and how, in terms of emission reductions. Yet, it was considered an improvement on earlier soft law instruments, by the institutionalization of follow-up mechanisms[1] (Shapovalova, 2016). The director of the Arctic Athabaskan Council, Michael Stickman, described the commitment by states through the initiative as the most important achievement during the Canadian chairmanship.

"Why do I say this when it is only a Framework Agreement lacking legal obligations?" he rhetorically asked, providing an answer himself: "because as a symbolical value, it may push normative frontiers on what is considered to be necessary environmental protective measures" (2015).

The negotiation process[2] in TFBCM was one characterized by 'high ambitions but low expectations,' meaning, political resistance was foreseen to stall a process dedicated to the solution of an environmental problem of magnitude. As the task force process reached its end, expectations turned out to have been unnecessary low: although many delegations would have wished for more, they yet considered the result satisfactory enough. This chapter follows the same structure and ambition as in previous chapter and is devoted to discussions on: (1) common understanding; (2) arguing; and (3) role learning. More concretely, and overall, it will show that the actors, using arguing as a means, managed to increase the level of understanding between them. By the same token, the Arctic states found themselves susceptible to (others') expectations. Just like previous chapter, it aims to show that expectations constitute a formative dimension of state behavior, which implies the behavior of states to potentially be molded – learned – through cooperative processes. However, also like the previous chapter, it will start with a background on the climate issue and how it found its way onto the AC agenda.

### *A Background on the Arctic Council's Climate Change Work*

When John Kerry urged the Arctic states to do more on climate change, this was a rhetoric known from before. Already in 1988, when campaigning for the U.S. presidency, George W. Bush Sr told the Boston audience that "those who think we're powerless to do something about the greenhouse effect are forgetting about the White House effect" (Hecht and Tirpak, 1995:383). In 1990, the Intergovernmental Panel on Climate Change (IPCC) released its first scientific assessment report, pointing toward human activities as having an enhanced effect on rising global-average mean temperatures,[3] quickly followed by the UN Framework Convention on Climate Change (UNFCCC) that turned climate change into a legal regime. The U.S. was the first country to sign in 1992, and by 1994 all Arctic states had ratified as well.

Within the Arctic context, the 1990s' global rise in climate attention was not reflected in the governance structure. The AEPS mentioned Arctic climate change, but only in terms of its significance for the global climate (AEPS, 1991: principle 2.2 iiic). Neither did the first AC Ministerial meeting mention climate change with any real vigor[4] (see *Iqaluit Declaration*, 1998) despite the timely context of the Kyoto Protocol and binding emission reduction targets for developed countries. However, operating in the outskirts, the AMAP had worked on its own to produce an Environmental Assessment Report in 1998, where climate change was included. Science on Arctic climate change, AMAP was convinced, would fill an important knowledge gap in IPCC's work (Stone, 2015:206–207). The chair of AMAP at that time, David Stone, later described how the author[5] of the climate change section was owed a great debt since she made "the Arctic Council and the circumpolar science community to take notion" (Stone, 2015:207). As a result, the Arctic Council started to request more information, including ordering the ACIA, to a large extent, funded by the U.S. (ibid.).

Even prior to its release in 2004, the ACIA became a success in terms of Arctic showcasing: it offered the Arctic states a way to present an Arctic imagery to the rest of the world (SAO meeting Espoo, 2001). The Arctic states approached the region as a global common, highlighting the need for responsibility: 'the whole of the world ought to protect the Arctic in order to protect themselves,' so they argued. And the Arctic states succeeded in their dramaturgy; for instance, Dr Robert Corell, chair of the steering committee of the ACIA, appreciatively noted how the exhibition hall of the UN World Summit on Sustainable Development was "impossible to pass without seeing a reference to the Arctic" (SAO meeting Inari, 2002). From now on climate change was introduced as a standing AC priority issue, repeatedly noted as of grave Arctic concern. Since the ACIA, the Arctic states have encouraged Arctic scientific findings to be implemented under the UNFCCC (*Reykjavik Declaration*, 2004), approached greenhouse gases in a united manner through, for instance, statements to the UNFCCC Conference of the Parties (COP),[6] and started to take action through short-lived climate pollutants.

The ACIA could be argued to have awoken the Arctic political attention on environmental issues,[7] but since then numerous other scientific reports relating to the Arctic climate have been produced.[8] In parallel to an episteme gone more focused on Arctic climate change, the SAO meetings, AC Ministerial statements, and declarations also disclose a climate tone gone sharpened, identifying a pressing and increasing need for mitigation actions (see, for instance, *Fairbanks Declaration*, 2017). It has positioned itself as an integral part of both those two objectives being the founding pillars of the AC: environmental protection and sustainable development.

### *Black Carbon and Methane*

Short-Lived Climate Pollutants (SLCPs) entered the AC agenda in 2008, brought by the episteme (SAO meeting Svolvær, 2008). Now the AC was no longer a global climate enlightener, but a regional climate actor with action-oriented goals of its own. Although states feared that an introduction of short-lived climate pollutants would detract global focus away from carbon dioxide and increase the emission reduction burden on the Arctic states compared to the rest of the world, the SAOs were generally supportive (SAO meeting Kautokeino, 2008).[9] The following year, climate change moved up the political agenda to the point where it represented a threat against the Arctic, as well as global stability (see AC Ministerial statements, Tromsø 2009). It was explained how the success of humanity in addressing climate change lingered on the knowledge of Arctic climate dynamics and causes (AC Ministerial statement by Norway, 2009). As a response to this, the SAOs recommended the ministers to establish a task force, which should draw on the work of AMAP and other expertise, and through the *Tromsø Declaration* in 2009 the AC soon initiated the task of seeking to find measures and actions leading to emission reductions on short-lived climate pollutants. In AC, such (political) focus has been on black carbon and methane.

*Methane* is a short-lived climate pollutant, with a life-length of about nine years. As a greenhouse gas, it has a far greater heat-trapping power than carbon dioxide. Methane is approximated to be the second biggest contributor to global warming, only outnumbered by – long after – carbon dioxide (AMAP, 2015). The Arctic is a region highly exposed to methane emissions in several different ways. Firstly, since methane is the main component in natural gas, it is a lucrative energy source drilled for and extracted in the Arctic or in its vicinity. When extracted and transported, leakages can, and do, occur. Secondly, since methane enhances the process where organic material becomes decomposed, it is also a bi-product in oil extraction. For oil rig operators, methane causes trouble, being highly explosive. Often, rather than to restore or cap, the gas is vented. It could also be lit on fire – flared – turned into less potent carbon dioxide, and soot. Thirdly, methane is trapped in the permafrost and the frozen seabed. Should the Arctic temperature rise high enough, there is a potential risk of enormous amounts of methane getting released. And fourthly, agriculture, food waste, and sewage sludge are other sources generating methane emissions in the Arctic (ibid.; Task force on SLCF, 2011; 2013)

*Black carbon* is the other short-lived climate pollutant of high interest in the AC, not to say of highest interest.[10] This is not a gas, but a particle shaped as soot. Being black, it absorbs heat, and therefore contributes to a general global warming as well as a specific Arctic ice-melting (the Albedo-effect). It is an extremely potent particle, creating a warming result that is much bigger than its size would imply, which also has a negative impact on people's health. Black carbon is derived from incomplete and inadequate combustion of fossil fuels: it could be generated from diesel engines running ships, from long-distance trucks and cars, and from wood burners and saunas. It is also generated when methane is flared, when wildfires occur, or when volcanoes erupt.

Now, it is not that all emissions from black carbon (or methane) found in the Arctic actually are emitted by the Arctic states themselves. Ship-owning businesses with home addresses far away from the Arctic, for instance, are sailing the Arctic waters, and particles are additionally travelling with ocean and wind currents from far away – for instance, heavy coal burning industries in China and elsewhere. Nonetheless, it has been established that emission-releasing sources within the Arctic nations have a greater warming impact, due to a higher effect per unit of emissions (AMAP, 2015; Task Force on SLCF, 2011; 2013:5).

### *How a Warming Trend Could Be Slowed Down: Suggestions on Action*

The TFBCM was not the first task force mandated to work on short-lived climate pollutants. The first one was led by the U.S. and Norway between 2009 and 2011 and had a focus on the technologies and economics of black carbon reductions. An expected increase in marine shipping qualified this source together with gas flaring as of special importance in the Arctic context. (Task Force on SLCF, 2011). The scientific evidence of the causal connection between the black carbon and Arctic warming seemed compelling, and Sweden's SAO was cited in *The Washington Post*: "The Arctic Council is very much about sustainability. Every country, they have to do their homework on black carbon" (Warrick and Eilperin, 2011).

Still, in 2011, the Arctic states were not ready for a common approach on black carbon reductions. Some SAOs voiced concerns over the task force recommendations to develop emission inventories and replace older vehicles and equipment with new technology having more efficient combustion. The SAOs found too little considerations to have been devoted to all of those different economic and regulatory realities that the Arctic states faced (SAO Meeting Copenhagen, 2011). More information was necessary, and the task force was asked to continue its work for a second run. With the key messages intact from 2011, the second task force delivery in 2013 clarified that Arctic anthropogenic methane emissions are likely to impact human health and Arctic climate. The U.S., Canada, and Russia were confirmed as global top emitters, and emission projections for the Arctic in total also led in the direction of increased levels in the coming decades. A reason could be found in the oil and gas sector, which was listed as the largest anthropogenic source of methane. In total, 60–75 percent of all potential methane emission reductions were in the oil and gas sector, and recommendations therefore included such things as use of new equipment and improved operational practices within this sector, to foremost minimize the leaks. Instead of flaring, which produces large quantities of soot, the task force encouraged states to either recover the gas (use it or sell on the market), or have it re-injected into underground reservoirs, so called cap and store (Task force on SLCF, 2013:7–9).

The articulated recommendations from the task force on SLCF, both in 2011 and 2013, had pointed the way forward. But recommendations were not in their own right – or standing – political. As a political response to the Arctic climate change, the Swedish invited the Arctic states' environmental ministers to a meeting in Jukkasjärvi, in Spring 2013. The ambition was to have the environmental dimension strengthened in the Arctic Council, not least in relation to a perceived necessity of urgent and decisive actions on emission reductions (Ministry of Environment Sweden, 2013). At the Kiruna Ministerial that later followed, the recommendation from the environmental ministers on political action was adhered. The next part of this chapter will discuss how the frame for political AC action was set up through the TFBCM process.

## Negotiating Emission Reductions: High Ambitions, Lower Expectations

Between September 2013 and November 2014, the TFBCM gathered for six meetings. One of the co-chairs retrospectively said: "It has been rewarding to see how all states – U.S., Russia, Canada, Norway, Denmark Finland, Iceland and Sweden, determinedly worked towards a concrete result ... [n]ot least have we promised to set up emission reductions target" (Kahn, 2014). At the same time, it was acknowledged how opinions had been many and the uncertainty big regarding the possibility to succeed within the timeframes (ibid.). For reasons like this, the mandate upon which the TFBCM originally rested had also taken height for a certain amount of resistance, using careful wording that would not cause any fright of joining in (interview 150318c). In hindsight, given what the task force achieved – an institutionalized structure around emission reductions – initial expectations on what would be possible to achieve turned out as unwarranted low. Or, with the words from a representative from the AAC, the framework ended up "less than we would have hoped for, but more than many expected" (Rosenthal, 2015).

The remaining part of this chapter is devoted to TFBCM and its consensus-reaching process on reductions of emissions from black carbon and methane, applying a micro perspective. Whereas common understanding prevailed in some respects, the chapter will also discuss how actors were expecting to disagree and therefore relied on a means such as arguing to reach a consensus. The last part of the chapter is set to illustrate processes of social learning, like those made vivid in arguing, by using roles.

### *Common Understanding on Soot, as well as 'Staying Put'*

#### *The Need for Reduced Particles of Black Carbon*

TFBCM and its different state delegations were united in two different ways: from the perception of it being valuable – for various reasons – to reduce Arctic emissions, and from the perception of Arctic relations as requiring good Arctic cooperation to thrive. To start with the former, as a starting signal for the TFBCM's establishment, the SAOs had confirmed the reduction of short-lived climate pollutants to be an AC priority, and informed the Ministers in writing:

> SAOs have considered *the appropriateness* [emphasis added] of moving forward on SLCP to bolster efforts to achieve substantial SLCP emissions reductions and encourage information sharing and recommend that the Ministers establish a Task Force to develop arrangements on actions to achieve enhanced reductions of black carbon and, in some cases, methane emissions in the Arctic, and report at the next Arctic Council Ministerial meeting in 2015.
>
> (SAO Report to Ministers, Kiruna 2013)

That substantial emission reductions are connected the wording 'appropriateness' indicates a somewhat changed normative context, where inaction on emissions no longer pass as legitimate. Reductions now come to result in an understanding of the inappropriateness thereof. As a delegate explained, a significant increase in knowledge had occurred over just the past five years, culminating in the establishment of the TFBCM, not least due to all the work conducted by previous task force(s) as well as the AMAP organization (interview 150623b). The time was ready to turn this knowledge into cooperation with a political focus, guided by science (TFBCM Two-pager, Whitehorse, Stockholm, 2013).

The first TFBCM meeting made it apparent that the participants shared the view on emission reductions as important for Arctic communities. The meeting was described as friction-free: "[w]e did not even have to use all of the time that was dedicated, which is how concerted the understanding was" (interview 150320). Also, at the last and final meeting, delegations stood unanimous in an "ambitious, politically aspirational collective vision" on emission reductions (TFBCM Two-pager, Tromsø, 2014). The two co-chairs were pleased with the result, a result that was described to derive from the commitment and ambition of all countries (Jacovella and Kahn, 2015).[11] When this commitment is elaborated further, what actors had in agreement, referral is made to a common understanding:

> [we agreed due to] the pretty obvious acknowledgement that what we were trying to do made a lot of sense [laughing]. It was obvious that we needed better information on black carbon in the Arctic, and it was obvious that all of the countries should be showing the Arctic Council what they were doing, in terms of mitigation measures for black carbon. So, there was a general buying of the concept behind the Framework, that we were addressing an obvious need and not doing something that would have little value.
>
> (Adviser 8)

Following Habermas, a good starting point for cooperation would be to establish 'what we know.' Thus, science on the need for reduced pollution, predominantly black carbon, had now passed as 'conventional knowledge'; the actors accepted this knowledge as being objectively true. As one delegate explained, "sometimes a task force is needed to establish what the facts are" (interview 150623b). In TFBCM, a common understanding was identified where one saw the need for emission reductions as 'a fact': "everyone agreed that something had to be done on black carbon and methane" (interview 141209). However, not everyone agreed to the reasons of *why* such action should be desirable: to reduce especially black carbon would be beneficial for more reasons than those pertaining to Arctic warming, especially to human health.[12]

Some countries do not primarily engage in this for the climate effect, but for the health effect. That's not less good, it's just as good. We find new ways to cooperate – different reasons to cooperate. That was a signal for us to move forward, that we knew – in the Arctic and in other contexts – that this was an area where we had a bit different interests but with the common interest of us all wanting to do something about it.

(Negotiator 2)

One delegate expressed a consensual disapproval concerning science, saying: ["t]he idea to form this task force was that science has been more or less established, done. I personally do not believe that science is finished within this area" (interview 151222). This means that actors shared a common understanding on emission reduction without necessarily confirming all science – *climate* science – to be correct. Following Habermas (1984), a generous approach to a certain issue uncovers more opportunities for states to justify actions on emissions reductions, given their respective social world (i.e., national context). To justify environmental protection could, for instance, more easily be done if it would be linked with health impacts or economic savings (interview 150611). Ultimately, the fact that soot entailed different benefits if reduced was understood to spur cooperation.

### The Importance of Being Cooperatively 'Stayed Put'

The negotiation process enhanced a dialogue amongst the Arctic countries. One delegate described it to have unpacked "a greater will to think collectively, to do things together" (interview 150318c). Another delegate described how it was "very constructive, in good spirit compared to climate negotiations" (interview 150521c). A willingness to cooperate should also have impacted on how the task force evolved, where the wish to 'stay put' superseded the Russian conflict with Ukraine. The TFBCM found itself in the midst of diplomatic turbulence when several delegations chose to boycott the third TFBCM meeting, held in Moscow, due to the Russian invasion of Crimea.[13] After the meeting, the Swedish co-chair reported: "Moscow was warm, although with an atmosphere that was affected by the situation between Russia and Ukraine." Another delegate explained the situation prior to the meeting as colored by diplomatic incidents: "the U.S. State department phoned the Canadian Foreign Affairs who phoned the Swedish Foreign Ministry...so no, the Moscow meeting did not turn out well" (interview 141012b). But states did not withdraw from negotiations:

It [Crimea] certainly did not help the process. Quite the contrary [...]. But, we still have a process and we managed to keep a dialogue throughout. After all, Russia *wanted* to commit in the Arctic issues together with the other Arctic states.

(Negotiator 2)

Therefore, states may not necessarily have trusted Russia per se, but they trusted a certain level of commitment in relation to black carbon and methane. In Moscow, Russia was also confirmed to be supportive of engaging in the TFBCM, and on a general level, a sense of feeling appeared similar to the following: "[a] good thing in this process was that we all wanted to reach a consensus, and *accept* consensus" (interview 150521c). Communicative action in the TFBCM seemed headed toward the values of Arctic cooperation, in addition to concrete values associated with emission reductions.

## Argue to Achieve! Commitment, Sector Reductions and AC Dramaturgy

Jokingly, a negotiator described arguing in TFBCM as "negotiations are not always smooth, and one should not get the picture that it was all angels with their harps, singing" (interview 151106). That hurdles could arise was also an expected possibility to be counted for. As previously discussed, arguing did not center on the importance of emission reductions per se, but on *how* to do it, and *what* to include. Predominantly, two things were argued about: (1) level of (national) commitment and (2) how reductions relate to certain emission-generating sectors. A third reason can be added as well, which sort of manifests itself as an overarching discussion on whether or not there would be a political value of telling the story – send a political message – of the Arctic region as highly cooperative on short-lived climate pollutants.

### Arguing on the Level of Commitment

The task force struggled with the issue of commitment in two different ways: how emission reductions should be regulated and distributed amongst members, and how to get everyone to accept the costs associated with taking action on emission cuts. Expectations on what would be possible to achieve were dampened due to the difference found in the Arctic states' social realities and political positions. With different capabilities, territorial sizes, driving forces, and involved stakeholders, an essential divide between the delegations concerned how far an emission reduction target should stretch in terms of commitment. One delegate described the political positions as rather locked:

> It becomes purely political, where one at the end punches each other's heads, arguing "don't you understand what I'm saying?" Norway says, "let's do even more!"; Denmark says "no we won't"; and Russia is lying in wait, saying "no, we cannot do more than we already do, otherwise we will not be part."
>
> (Negotiator 6)

Most countries, including Finland, Norway, and Sweden, strongly advocated the idea of having a quantitative target, where one should agree to a specified figure equivalent to what in percentages should be reduced, for instance, a 50 percent reduction by year X (TFBCM Two-pager, Iqaluit, 2014). One HoD motivated this by reference to the importance of explicit expectations:

> It was many many years ago that politicians said that nitrogen emissions [from shipping] had to be reduced by 50 percent.[14] Scientists thought it was a terrible mistake because we did not know if it would be achievable. But after 30 years, it has been reached. [...] It is like a guiding thing. For example, if we have agreed on numbers in international processes – it is not mandatory, it is not binding, but we have still said to obey it – then on the national level, we can justify the means. In relation to leaking, or this sauna, or whatever we [as states] have. This was the idea: it is good to have numbers if you are trying to achieve something.
>
> (Negotiator 5)

Some Nordic states even brought up the need for *national* targets on emission reduction, but such an ambition these states soon had to abandon (interview 150320). However, also a quantitative target was considered too difficult, as being premature; not all states had their emission inventories ready enough to be connected to specific goals. At the last meeting a consensus was finally reached on a "collective quantitative aspirational goal," although it was not set to be decided until 2017. The result was described as a pragmatic leveling between those who felt uncomfortable with targets and specified obligations and those who strongly advocated national actions on emission reductions. The result also represented a plan forward: "[w]e have created a vision about a target, and doing that we are pushing politics further although one is not committed to – in concrete terms – do this or that" (interview 150318c).

Cooperation in the Arctic Council is cooperation highly respectful of sovereignty. Acknowledging this, the TFBCM gave expression for a negotiation strategy where one sought to bypass political resistance by searching to attain small steps rather than ultimate outcomes; step-by-step expectations on cooperative behavior (i.e., emission reductions) would be integrated into prescriptions on appropriate AC behavior. Thus, to agree on information sharing over borders and to develop emission inventories would not in itself reduce emissions, but as illustrated by the following quote, would be a step on the way toward reductions in the future:

> If you are reporting your pollution levels you will have an incentive [small laughter] to reduce those pollution levels [...] I think we knew early on that negotiating a legally binding agreement would not be the

right first step for the Arctic Council. It would have been maybe too
much too soon, and it could have been a lost opportunity to create this
very basic but very important reporting function that we now have.

(Adviser 8)

To have arrived at such a consensus where a target was accepted meant
that the Arctic states also were ready to accept certain costs that come
attached. That the 'Arctic majors,' such as the U.S., Canada, and Russia,
employed a greater will to join the Nordic countries in their climate
ambitions – and showed a greater flexibility than one would have reason
to believe – was understood in the task force as progress. Although Russia
was not the only state expected to reveal a resistance to (too demanding)
commitments, it yet was a constant concern echoing amongst delegations:
"the biggest and most difficult task was to know whether the Russians were
on board" (interview 141210b). The concern had its origin, at least partly,
in past experiences from previous task force(s), which now caused doubts
also in TFBCM "regarding how far a certain state would stretch in order
to commit to anything at all" (interview 150318c). Therefore, the reached
agreement was described to be new compared to previous task force(s):

> [T]he dynamic changed, and there were no significant voices going
> against suggestions on closer cooperation. No vetoes against entering
> the "new room" [...] Nobody was opposed, and that was a new lan-
> guage. That is silence; silence was a new language.
>
> (Negotiator 1)

Through arguing, a more cooperative language arose and the Arctic states
managed to overcome some of those problems associated with cooperation
on emission reductions, like differing social and subjective interpretations
thereof. Following Habermas (1984), it is furthermore an illustration of communicative action that is oriented toward enhanced understanding – of each
other as well as the cooperative topic. To increase the understanding of each
other would be a driving motivator for all negotiations, explained by one
of the participants in the following wording: "the sole basis for each negotiation is to understand the opponent's wants. Otherwise, negotiations will
stand still. Therefore, one must learn why a certain actor is doing a certain
thing" (interview 141209). In the TFBCM, arguing generally led to improved
communication between participating actors, which resulted in a better
understanding of each other. As such, it seems that some of this interaction
regarding emission reductions was leading toward greater commitments in
cooperation, than initial expectations would have suggested.

### Arguing on Sector Reductions

The TFBCM aimed to derive more information regarding the levels
of Arctic emissions, and in general passed discussions on how emission

reductions relate to specific national sectors. At this stage, as the Framework Agreement later came to specify, more was expected to follow once the agreement was adopted by the Ministers. In relation to black carbon, in particular, joint decision on action was believed to be restrained for three different reasons, pertaining to the Arctic states' national circumstances: a lack of statistics and common measurements (on foremost wildfires and gas flaring); cultural habits; and economical costs. Finland and Sweden, for instance, described how a cultural habit like that of wood combustion would be the issue to tackle for the coming five to ten years, which is not being without difficulties: "[i]t is a very challenging sector because it goes to peoples,' somehow, everyday life. And there are strong feelings" (interview 150521c). For other states, a lack of statistics had an effect, as well as costs, where some pointed toward the Arctic states entering the field of emission reduction cooperation unequally:

> The greater territory you have, the more problems you will have with such things [emission inventories and regulatory reports]. If you have a minor territory with a single source of black carbon, you can easily describe it and you can easily include it in other pictures. But when half of the picture [laughing] is your responsibility, then you are in more trouble than others.
>
> (Adviser 10)

That costs mattered was also noticed in a sector such as Arctic oil and gas being left out from discussions. It had been brought up by previous task force(s), since it represented the largest source of anthropogenic methane emissions. Through flaring, it also represented a significant source of black carbon particles (Task force on SLCF, 2013:5–6). But in the TFBCM, the close relationship between gas flaring and coastal states' energy production was considered a particularly aggravating economic circumstance for achieving emission reductions:

> One should know, some of the major sources relates to the energy system: production, distribution, and of course combustion. But if we think of production, then quite some black carbon emissions stem from flaring. And…this has been a bit difficult. Partly due to us having too little knowledge about these emissions, but also because they are tightly associated with gas extraction.
>
> (Negotiator 2)

To decrease the usage of flaring as a neutralizer of explosive gas requires energy stakeholders to be involved. And it is furthermore costly: "to reduce flaring would take investments, which are not always welcome" (interview 160623b). However, in the TFBCM, talk on sectors was – as previously mentioned – redundant, left out from discussions. By the same token, this sector was kept restrained from causing friction amongst delegations.

Delegates talked about negotiations that were not concrete enough, where it was avoided to speak about who should do what and when, and where visions rather than practice were the guiding torch. One HoD identified oil and gas interests to form a hidden agenda. As something that had an influence on the final paper but without being openly discussed (interview, Negotiator 5). From a social perspective, arriving at emission reductions was thus more difficult than acknowledging it as relevant in relation to the objective world. And yet, if just allowing for time to progress, another delegate explains how gas flaring was a topic that advanced on the agenda:

> [i]t is not that anyone has a closed discussion, put the brakes on the task force work [on flaring]. Instead it has now been prioritized as perhaps the most important issue – or at least *the first* issue to deal with, perhaps not the most important one.
>
> (Interview 150623b)

The quote signals states to be attentive to environmental norms, even though these go against material interests, as long as they have the time on their side.

### *Arguing on the AC Dramaturgy*

A third theme of arguing concerned whether TFBCM was too political in its ambitions. Some delegations questioned what was interpreted as being 'speedy conclusions' that some other states, acting as environmental entrepreneurs, relied on in their preferences for far-reaching targets and commitments. These states were understood to use politics as a normative driver, instead of letting science be prerogative where more knowledge was needed in order to see the full picture of pollutants' cause and effect (TFBCM Two-pager, Whitehorse, Stockholm, 2013). A quote illustrating one such voice where the process was understood as too political is the following:

> Those who worked with the Arctic were mostly politically focused; they were focused on the political significance of what more we should do to reduce, of the cooperation as such. They were not interested in practical national politics, of having things implemented, and they had not focus on whether it would have a sound effect or co-benefit in other areas. It was just politically isolated, "let us have requirements on reductions of black carbon and methane."
>
> (Negotiator 6)

Indeed, especially those delegations who viewed themselves as being environmental norm entrepreneurs interpreted the mandate as an opportunity to push norms closer to environmental protection. In the same token, they deliberately tried to change notions on how far the norm on environmental responsibility would stretch. In that way, the TFBCM shared resemblances with 'theatrical acts' of state performances and spectacles that served to manifest political ideas as real (see Geertz, 1980). However, not only were these political ideas directed at Arctic relations, but 'orchestrated spectacles' were also directed at an external audience outside of the Arctic, and here the Arctic states shared more of a common understanding. Previous task force(s), for instance, had identified black carbon and methane emission reductions to offer the AC an important leadership opportunity, where a "common Arctic voice" could convey and demonstrate to external actors and observer states the importance of Arctic climate protection (Task force SLCF, 2011:10, 2013:13). Arriving at the TFBCM meeting, the Arctic states were unanimous in their support of the AC taking regional leadership on emission reductions, while also sending a "strong political signal" to international climate negotiations on the need for action-taking (TFBCM Two-pager, Tromsø, 2014).

Similarly, to create a solid base where observers could join in beneficially was a high priority for many delegations. The last task force meeting was partly devoted to the issue of motivating observer countries to join in: "the door is open and we hope for really good cooperation, but how do we do that and how do we convince others that it is a good thing?" (interview 150521c). A start, one could assume, would be to present a consensus-oriented Arctic dramaturgy on emission reductions, where the Arctic states all seemed dedicated to the task.

*Summing It Up*

So far, this chapter has relied on a micro perspective to analyze negotiations as a consensus-reaching process. Through this perspective, it has identified where states share a common understanding on the topic being negotiated, as well as where political resistance toward change, i.e., enhanced environmental governance, seem to be the most predominant. In the following section of this chapter, the fact that some states and state representatives are calling for change whereas others attempt to resist will be linked to a discussion on states' roles. More precisely, those positions obtained by states in the negotiation will be discussed relying on Mead's distinction between *I* and *Me* as together forming part of an actor's self-consciousness. As such, this last part of the chapter will illustrate how role performances are elaborated in relation to alter expectations, which at least partly signals others to influence role-playing and cooperative outcomes.

## Continuity and Change in Role Performances

In TFBCM, those with the high(est) ambitions did not get to see all of their ambitions being realized, but they did not get to see them being abandoned either. It has previously been suggested that a reason for why many delegates had low expectations had to do with the level of resistance – economic, practical, and cultural – that cooperation on emission reductions expectedly would activate. These low expectations prevailed also after the work of the TFBCM had been complete, where delegates expressed doubts on whether the agreement really would be signed by the Ministers. However, in the following, it will be argued that these quite low expectations of environmental commitments and, ultimately, Arctic relations, failed to fully consider the impact of alter and actors' tendency to follow role prescriptions.

Roles within the TFBCM followed the role pattern as described in Chapter 4. Compared to the TFOPP, states still activated different dimensions of their roles. For instance, whereas Russia was vocal and informative in the TFOPP, they were silent in the TFBCM; whereas Sweden and Finland held low positions in the TFOPP, they were active and outspoken in the TFBCM; whereas Denmark was a kid-brother emphasizing cooperation in the TFOPP, they became the realist who above all searched for pragmatic environmental efficiency in the TFBCM. Furthermore, this time it was not coastal states that acted as the leading nations, but environmental norm entrepreneurs, suggesting the issue of black carbon and methane to be different in character compared to oil spill prevention, with different expectations attached concerning roles and role performances.

### When the Role Is Confirmed: The Leaders – Norway, Sweden, and the U.S.

*Norway.* In the TFBCM, Norway was a leader who had farfetched ambitions for the cooperation. An important issue was to make sure that methane stayed in the agreement. A negotiator from another Nordic country explains how this greenhouse gas was met with resistance: "I understood it was too early. […] They said very clearly, that this is…they can accept black carbon but they cannot accept methane" (interview 150611). Compared to black carbon, a shared understanding on this climate forcer was less obvious, less held in common (interview 150623b). However, with methane being estimated as the second most important greenhouse gas after carbon dioxide, and where the Arctic states were responsible for about 20 percent of global emissions,[15] Norway was in the frontline regarding methane's inclusion in the agreement, preferably with reductions attached.

'The focus on methane had, though, nothing to do with Norway as a petroleum nation,' a Norwegian state representative jokingly explained. This, however, would be a truth with modification. Whereas the words illustrate Norway to be well familiar with its duality in roles (environmental protector/petroleum nation), it could be argued that Norway's legitimacy as a petroleum nation presupposes the playing of also an environmental pro-active role. Norway combines them into being an 'environmental petroleum nation,' noticed in things such as: a long ban on routine gas flaring, financial support for the World Bank's initiative to eliminate routine flaring (amongst others) by 2030, and decreased levels of methane emission ever since 1990 (Norway's National Report on Black Carbon and Methane, 2015; Government of Norway, 2015; Carbon Limits, 2014:28–29). To Norway, who has its ego built around both petroleum and environment, taking strong actions within a task force on short-lived climate pollutants, is in line with how its inner and impulsive *I* would act. The rest of the TFBCM participants also recognized Norway as such an active and progressive environmental leader, socially reinforcing the Norwegian ego role conception in line with that.

*Sweden.* The TFBCM discussed issues that Sweden knows well from before. Therefore, everyone participating expected environmental entrepreneurs to push the agenda, providing Sweden with the mandate to express – and confirm – its role of being a teacher as well as a leader. Nationally, Sweden had come far regarding emission inventories, as well as real cuts in emissions of black carbon and methane (Sweden's National Report on Black Carbon and Methane, 2015; interviews), which is why it considered itself to be able to act as a trustworthy teacher. What is more, Sweden also felt that others had expectations of them to take on such a role:

> There were expectations on us leading and pushing the process forward, absolutely. But there were times where we moderated to let others – or to *get* others…the negotiation dynamic will not be good if one is pushing too hard. There must be a possibility for reaching an agreement as well.
>
> (Swedish delegate, 2015)

That a small, otherwise often silent, non-coastal state like Sweden took on a leadership role in the TFBCM was still in line with role continuity, and not change. Climate change, and not least short-lived climate pollutants, is an area where Sweden has good national command and international reputation. "Some countries were in better shape," a delegate recalled, "I remember Sweden was pretty – how to say – aggressive and ready for many things" (interview 151222). In the TFBCM, it was not only that others expected Sweden to lead and push, but that the TFBCM mandate called for progression on short-lived climate pollutants. To take on the lead and be in the forefront regarding emission commitments would therefore be highly appropriate given the interactive structure, and in line with Sweden's role conception. Indeed, together with Norway, Sweden was described as a "very vocal participant" (interview 151106). That there is a cohesion in expectations between ego and alter, between *I* and *Me*, is important for Sweden when attempting to exert social power, since it provides a social context ready for a teacher and ready for norm entrepreneurial behavior. Thus, neither Sweden nor Norway had to adapt their role performance, in terms of a leveling between ego and alter, since these rather had each other confirmed.

*U.S.* The U.S. also confirmed ego through alter, *I* through *Me*. In Chapter 4, it was discussed how the U.S. was more of an innate leader than an 'Arctic' leader with clear Arctic goals and interests set out. The first half of its AC membership was carried out a bit reluctantly, which also included climate change work in general and ACIA in particular, which for long was held in contingency (Nilsson, 2007; Nord, 2016). However, from 2009 onward, the U.S. revealed a change in attitude toward the AC cooperation, which was understood as key to responding to the Arctic challenges (Pedersen, 2012). Such challenges included the climate, and the change in attitude was timely related to becoming a leading actor within the area of short-lived climate pollutants. Within the TFBCM, a delegate describes the U.S. role as being both an innate leader, as well as being climate aware:

> [A state that was] very active, which also put a lot of resources in work, was the U.S. Its delegation was very good, also in developing the document toward a good aim and end. They were listening closely to what the other countries wanted to express, and they came up with constructive solutions to the text, which incorporated those aspects.
>
> (Adviser 7)

However, some noticed the activity level to increase after negotiations had continued for a while. A delegate describes how one had been ready to give up on quantitative targets due to tough discussions with the U.S., when the country changed position: "they could agree to quantitative targets of some kind. From having opposed such a discussion, they became a leading actor in the same discussion" (interview 150320). Others pointed to a leadership based upon its previous co-chairing of the task force(s) on short-lived climate pollutants (2009–2013), and its good scientific expertise.

The fact that the U.S. was up next concerning the Arctic chairmanship, references were made to the U.S. as planning to administer cooperation on short-lived climate pollutants during its chairmanship, keen on being able to present some good solutions when hosting the Ministerial meeting in 2017. Delegates also noticed the U.S. as displaying a new level of climate commitments in its international politics outside of the Arctic. For the U.S., it seems like its ego *I* – which views itself as a leader – had to meet its alter *Me* who directed light on to environmental norms as being on the rise in the Arctic. Thus, these norms acted as prescriptions on how a true Arctic leader should act. In that sense, the U.S. confirmed its inner *I* by adhering to its social *Me*.

### A Problematic Situation Admitting for Little Learning

The Danish AC role resembles a pragmatic (coastal state) kid-brother. However, due to Denmark's good track record regarding climate change and airborne pollutants, there was no need to include the 'kid-brother' in the TFBCM. Instead Denmark chose to disregard much of the alter expectations, since these were believed to move the TFBCM process in the wrong direction – a direction that would not generate the most efficient emission reductions. As such, the role performance of Denmark resembles a pragmatic agent-driven behavior, where actors are less keen on letting others exercise influence over the chosen behavior. A reason could be found in Denmark's ego expectations, which rely on environmental pro-activeness, not least in climate issues, and provide Denmark with confidence – a confidence in knowing that others also view Denmark in such a way. Indeed, other delegations described the state to be dedicated to matters of fact, not visionary but keen on arriving at a good agreement that would make a difference.

Denmark was concerned over climate efficiency: was there not a risk that the TFBCM would duplicate already established regimes, and did the Arctic states really know all there was to know about black carbon and how it behaves when transported over long and short distances alike? It also expressed concerns over the risk for free-riding and flawed analysis regarding what emission reductions targets would do for each state's competitive advantage. As such, Denmark did not appreciate the political take on the TFBCM, which failed to keep in mind *efficient* emission reductions and what emission reductions would mean, in terms of economical/practical consequences for each state. These concerns could all be connected to the Danish role of chairing environmental protection and climate work, while also keeping sovereignty and integrity high on the agenda. The fact that not all the Arctic states were expected to contribute to emission reductions was thus considered problematic.

This could as well provide some understanding on Denmark's reluctance to high ambition levels on methane reductions, as advocated for by Norway, for instance. For a state that has done a lot on climate change and expects to achieve considerably higher percentages of emission reductions than the EU average (Danish Ministry of Energy, Utilities and Climate, 2018), further targets on methane emissions was explained by the Danish delegation to have little left to target than agriculture, and then more precisely, a reduction in livestock – the number of cows. The absolute vast majority of methane emissions in Denmark stem from agriculture and cattle; furthermore, Denmark expects an increase in animals for the years to come (Denmark's National Report on Black Carbon and Methane Emissions, 2015:9). At the same time, Denmark is already working, according to an EU regulation, to achieve a 40 percent reduction in emissions from agriculture by 2030 (Hansen, 2016). To an agricultural nation like Denmark, the cultural and economic value that is protected through having livestock preserved would be considerable.

Thus, whereas other environmentally proactive states, to some extent, had their hopes high on Denmark joining the Nordic states in their attempts to reach ambitious targets, the Danish position proved impervious to any such expectations, despite 'matey' appeals of acting as supportive, or an expressed surprise over what was considered a Danish retrogressive behavior (interviews, 150318c; 150611; 150612). Denmark's immunity to alter expectations could here be understood in relation to its quite positive track record concerning climate commitment, where its genuine interest in improved emission situation for the Arctic was also understood to be known by others. Denmark's social *Me* was therefore less concerned with prescriptions on environmental norms, and more devoted to its role character of being pragmatic and a realist. The Danish environmental conscious is – one could assume – considered to be clean and expected to be recognized by others, which is why Denmark could take on a more pragmatic role in the TFBCM, centered on efficiency and (national) fairness, without having to risk their credibility becoming lost to the climate issue per se.

### Feeling Misunderstood? Two Territorial Giants Reconsidered

Both Canada and Russia, the two states with the biggest territory to monitor, were viewed as dragging their feet a bit, whose level of commitment was unsure, and who seemed to find support in each other. "It was often that Russia voiced its opinion in favor of Canadian arguments," a delegate said, "although it felt unscripted" (interview 141210b). However, as participants in the TFBCM, they were not immune from alter expectations revealing them as being less committed.

*Canada.* As shown in Chapter 4, the ego part of the Canadian role conception includes two dimensions pertaining to protection: safeguarding the region through value-laden stewardship and ensuring self-determination in order to stand firm against external pressure on sovereignty. These two components also echoed when France Jacovella, co-chair of the TFBCM, in content wording declared the TFBCM result to be a result with which they were pleased:

> The Task Force's outcome signals the commitment of the Arctic Council to lead in taking enhanced action to reduce emissions of black carbon and methane within its borders. At the same time, the Task Force signals that cooperation with partners outside of the Arctic region, including contributions by Arctic Council Observers, will be essential to tackle the issue effectively and reduce the impact on the Arctic and its peoples of black carbon and methane that is emitted beyond the borders of Arctic States.
>
> (F. Jacovella, Canada, co-chair TFBCM, 2015)[16]

Through the work conducted by the TFBCM, the AC signals to take the lead: i.e., Canada is thus acting as a steward protecting the region. Canada foresaw a document that would make a difference for communities on the ground, relevant from both a health perspective and an environmental perspective, and something that would be more than a report sitting on the shelf. The work on short-lived climate pollutants was declared a highly prioritized work for Canada and its government, a chance to come up with concrete measures that would benefit the North (TFBCM Two-pager, Whitehorse, 2013; interviews). And yet, this was not how other delegations understood the Canadian role performance; they rather saw a not very active state that attempted to resist. Responding to a question on whether Canada could notice any expectations from others regarding how to act, the answer touched upon a role conflict:

> We have always played a strong role in the Arctic Council initiatives. And whatever others might think of Canada in the climate change world, that is a different story. This initiative is not Canada negotiating global climate change. It is initiated by Canada being engaged in what we can do regionally through the Arctic Council to move the issue of short-lived climate pollutants along.
>
> (Canadian delegate, 2015)

Canada here seems to wrestle with others' expectations of the Canadian role, expectations they frame as misunderstandings. Canada is keen to separate its international climate behavior, for instance, its withdrawal from the Kyoto Protocol, from the Arctic stewardship behavior on short-lived climate pollutants, which is described to be 'something else.' Regarding the latter, Canada explains how the country has always been very supportive of the AC's work on short-lived climate pollutants. What the quote illustrates therefore seems to be an attempt by Canada to get others' expectations to move beyond perceptions of Canadian role performance on climate change and look to the bigger stewardship picture where Canada rather has a history – for others to see – of strong AC commitment.

In practice, or at least rhetorically, Canada also tried to prove expectations wrong, and described the Framework Agreement as a document to support the Arctic states to move forward on emission reductions, and where the Canadian chairmanship had showed leadership (AC Ministerial statement by Canada, 2015). However, the other delegations did not really share this view of Canadian role performance in TFBCM, and gave instead account of a Canada acting a bit retrogressively, whose comments were picked up upon by Russia, and who did not distinguish itself as environmentally proactive. "I don't know their reason for why they were trying to make this agreement so weak," a delegate expressed (interview 150611). The TFBCM took place in a context of Canadian AC chairmanship that – although successful in other areas – received criticism for its low priority of environmental issues, as well as its hard line with Russia (Exner-Pirot, 2016). The appointed chair, Leona Aglukkaq, also created some headlines when she, prior to the chairmanship period, proclaimed Arctic warming – although clarifications were made later – as debatable (De Souza, 2013). Therefore, it could be that the role conflict identified in the TFBCM was not only manifested in others' expectations, but, even more troublesome, was located in the Canadian *I*. Thus, although Canada is keen to be understood as a cooperator, *and* reacts toward alter expectations, incoherence in ego role conceptions makes it difficult to find solutions to please them all.

*Russia.* Amongst the different states, Russia would be explained as the one to cause the highest degree of uncertainty regarding commitment: "to negotiate with Russia is *always* difficult," a delegate with long AC experience mentioned (interview 2014). It would also be the state giving rise to most diverging understandings: "Russia did their best," one delegate said and attached positive connotations thereof; "Russia is lying await," another one stated, "ready to withdraw should anything be too demanding, too concrete." "Russia is silent," a third perspective went, interpreting a lack of thematic interest and poor political anchoring"; and "Russia is silent!" a fourth perspective noted, referring to a new negotiation climate of (normative) acceptance and political willingness.

This indicates not only a flawed understanding of Russia, but perhaps foremost, in general, a lack of subjective trust regarding the Russian character. The same interviewee who described negotiations with Russia to be difficult refers this to a different type of culture: a culture to some extent linked to a negotiation style, but also to understandings of Russia's interests and behavior in the Arctic; a lion share of the AC projects has, for instance, been dedicated to clean up Russian territory. At the same time, large-scale pollution has continued, for instance, through industries and pipes of low standard (Blokov, 2011; Greenpeace, 2017; WWF, 2014). In that sense, others' environmental expectations were low. As the TFBCM process went on, expectations did, however, go up:

> Russia was less prepared for this issue than other countries. They do not have an emission inventory yet. As a result, they have been more hesitant than others to commit. …Even committing to the idea of submitting a national inventory was not an easy decision for them. But they have come around, and I think they do want to be constructive, so I'm optimistic.
>
> (Adviser 8)

Retrospectively, one could establish the Russian role in the TFBCM to be a low voice performance, who for sure pushed the brakes by cherishing "thoughtfulness" and reminding others of the importance of having science to support any actions. But it also was subjected to political boycotts, yet it signed the agreement and also promptly handed the national inventory report in, as requested in the Framework Agreement. If Russia itself would provide an answer on why it was doing all this, the answer would be self-explanatory: "had the topic not been important to Russia, we would not have participated in this" (Interview, 2015).

If we accept this account, others' expectations would seem to have turned out – at least partly – wrong. They were wrong because they either failed to account for a change in attitude regarding Russia's perceived value of emission reductions, or because they miscalculated Russia's willingness to adapt to the interactive structure; most likely both. That Russia has a social *Me* that does not want to be too impossible to deal with is manifested in the negotiation strategy used by others, i.e., "having *everyone* participate in order to get Russia to join" (interview 150612). As such, it is acknowledged that Russia is not immune to alter expectations, that it can be persuaded and is somewhat accommodative, although it takes everyone else to contribute first. Thus, Russia, just like the other states, seems to care for its reputation. Such caring for reputations could also be the reason to why the Framework Agreement was given an open structure that is encouraging everyone to join in but who does not expose anyone failing on commitments. One delegate explains:

> [It will be] no Kyoto, and no one who should be able to say you have not fulfilled your obligations! But within those limits, one can yet create something more active than wishful thinking on doing the best we can.
>
> (Negotiator 1)

Russia is a storyteller, someone who wants to write the manuscript by presenting images of a grand Arctic state or a grand Arctic cooperation. Despite this, the not so vocal way that Russia acted in the TFBCM still fitted the picture: with a task force process described as disclosure *silence* as a new language, then a lowered voice becomes an expression for *Me* in a context where *I* have no clear informative role or national story to tell. It becomes the language of the 'black sheep' who battles with negative alter expectations, but whose ego still expects of itself to prove to be a cooperative Arctic state. For a state with major Arctic ambitions, one cannot afford losing respectability by failing an issue described as fateful to the Arctic future. For that reason, it may be that Russia, as the only state, uses the Ministerial meeting as a platform to inform about its national undertakings of preparing an emission inventory report, which it announces to be "part of fulfilling its international commitment under the Arctic Council" (AC Ministerial statement by Russia, 2015).

It could be that states who 'do not want to commit' generally are 'afraid to fail.' Although a change in words does not render states, like Russia, keener to accept emission reduction targets, it does present a state as being more susceptible toward social interaction and alter expectations. In Russia's case, it would suggest that others' low expectations on this state, mainly departing from prior experiences, did not fully consider the impact of interaction on state positions. Following this, a suggestion is that states acting more retrogressively in the TFBCM possibly are more considerate to structures prescribing the behavior of 'a good cooperator,' than would apply to climate proactive states. These latter states can afford a gamble on ambitious reduction targets since nothing would be forsaken should a failure occur: they would still be in the climate front lead with role conceptions preserved. For states being less proactive, a failure would potentially threaten their role conception in such a way that they would be confirmed as bad cooperators, i.e., as 'bad Arctic states.' A price possibly too high to pay.

## Conclusion: A Negotiation Process Attentive to Climate Prescriptions

In the beginning of this chapter, the negotiation process on short-lived climate pollutants was described as characterized by high ambitions regarding Arctic emission reductions, but lower expectations regarding what really would be possible to agree to. Low expectations could perhaps even be better explained as *careful* expectations: states did not want to put their hopes up too high. However, after the task force concluded its work, the Arctic Council stayed on course to deliver on emission reduction ambitions, where the Ministers adopted concrete goals of which the Arctic states committed themselves to collectively lower their black carbon emissions by at least 25–33 percent, below 2013 levels, by 2025. In that sense, the norm on climate change, i.e., Arctic short-lived climate pollutants, turned out as prescriptively rather strong, or at least, stronger than what delegates expected when taking into account political resistance – shaped as, for instance, state capabilities and national interests.

The common understanding that existed among the Arctic states in the TFBCM indicates a norm internalization to have occurred, where no state could afford to question the relevance of reduced emissions. One could question bits and pieces regarding emission reductions – hence, much of the TFBCM arguing – but one could not question the overall scientific message about these pollutants being bad for the Arctic climate *as well as* people's health. Therefore, to do as many of the Nordic countries did, and politically push for a quantitative and visionary target, could be a way to speed things up. Ambitions – even those which are not scientifically approved yet – there-through become part of the interactive structure, the theatrical setting, in such a way that others make their acquaintance. The possibility for socialization to occur arises, changing expectations on required type of behavior from a state claiming to be not just any state, but an Arctic state.

An essential argument of this book is that 'small' cognitive processes also contribute to an enhanced understanding of how change – real or potential – may come about in international relations. Therefore, just as previous chapter, this chapter also offered an entry into the domain of state learning through a micro-level analysis of one specific Arctic Council negotiation. Through such an analysis, the chapter has disclosed different interactive logics and understandings to impact differently on a state's willingness and ability to make strong emission reduction commitments, depending on whether the environmental issue – and the international cooperation thereof – is looked upon in an objective, social, or subjective way. By the same token, the micro perspective has revealed interpersonal interactions – for instance, perceived intentions and trustworthiness of others – and not only material state capabilities and interests, to have an influx on the development of state negotiations. In the same manner the micro perspective has given access to how states and state representatives come to reflect upon role suitability, given a contextually anchoring thereof.

For instance, with all of the Arctic states' roles being constructed around a notion of, in different ways, the protection and promotion of not only the Arctic but as themselves being *Arctic nations*, entailed is a role conception that relies on the ability to reach out to others and cooperate, should the region be safeguarded. Herein lies a difference between TFBCM and TFOPP (Chapter 5), in terms of norm prescriptions and subsequent role performance. Compared to oil spill prevention, the norm prescribing a reduction of short-lived climate pollutants has come farther in its internalization process, requiring states to act and to shoulder greater environmental responsibility. Whereas oil spill prevention is acknowledged as an important norm, its prescriptive qualities seem to be less clear and in more of a competing relationship with the norm of state sovereignty. The issues being negotiated in TFBCM and TFOPP, respectively, also carry different incitements for states' role performances. Different parts of the role get activated depending on the issue area that is being negotiated, which is telling of a role flexibility inherited in the way that states perform and socially interact with others.

This chapter has also, just like the previous chapter did, disclosed alter to affect roles, i.e., revealed *others* as mattering. However, the small case studies provided for suggest that those states that were acting as leaders, foremost Norway, Sweden, and the U.S., also were those who were the least affected by alter and others' expectations. A reason thereof would be the lack of friction both within and between these expectations and role prescriptions, thereby causing behavioral adaptations superfluous. To Denmark, Canada, and Russia, alter expectations brought afore a need for reflection regarding the (mis)match between ego and alter expectations, and how to best perform the role. For Canada and Russia, these reflections showed a sensitivity to the perceptions of others. This sensitivity pertains to a wish of being recognized as 'an Arctic cooperator,' which also offers input on how it may be that those states who are perceived as less environmentally proactive yet are not immune to alter expectations. Together with symbolic interactionism, the role theoretical perspective applied to Chapters 5 and 6 has turned out valuable by disclosing 'others' as constituting an integral part of a state's role, impossible to stay separated from but rather acts as a constant reminder of cues and demands to some extent to being greater than the state itself.

## Notes

1 States are obliged to in iterative two-year processes report on their levels of emissions and mitigation strategies (Framework Agreement, 2015).

2 The material for this chapter mainly draws on interviews, where 12 interviews have been conducted with TFBCM participants specifically. Iceland did not

actively participate in TFBCM, but the remaining 7 states are represented in interviews: with some delegation sizes being as small as one (1) representative, this means full interview representation in some cases.

3  In this first IPCC report, in 1990, the greenhouse effect caused by humans and the natural greenhouse effect was yet a relationship in need of further research. The causal linkage between climate change and anthropogenic greenhouse gas emissions was established in the fourth IPCC report, in 2007.

4  It welcomes the overview and assessment of the CAFF and AMAP regarding changes in the Arctic ecosystems, including the effect of climate change and UV-B radiation on these ecosystems (Iqaluit Declaration, 1998: paragraph 21).

5  Elizabeth C. Weatherhead.

6  At the World Summit on Sustainable Development (WSSD) in 2002, linkages between Arctic warming and global climate change was for the first time showcased. Similar messages have since then been sent at COP meetings, where the Arctic Council, for example, has put up Arctic booths for educational purpose (see, for instance, SAO meeting Anchorage, 2015; Fairbanks, 2015).

7  The ACIA is understood to represent a clear before and after, regarding environmental focus, which brought more political attention (and funding) to the AC (AC, 2004; interviews 2014–2015).

8  Amongst these are CAFF's *Arctic Biodiversity Assessment* (2013), PAME's *Arctic Marine Shipping Assessment* (2009), AMAP's *Arctic Oceans Acidification Assessment* (2013) as well as AMAP's *Snow, Water, Ice and Permafrost in the Arctic* (2011; 2017). In 2016, the *Arctic Resilience Report* was published, focusing on the importance of building resilience and slowing down Arctic change, justified for the sake of the Arctic as well as for the rest of the world.

9  For instance, the same year in 2008 in Stockholm, the Global Atmospheric Pollution Forum was held, in order to find ways to strengthen international cooperation on airborne polluters, where the climate benefits of reducing polluters with short lifetimes were highlighted. Later, in 2012, the Climate and Clean Air Coalition was created to reduce black carbon and methane. Three of the six founders were Arctic states: Canada, U.S., and Sweden (Stockholm Environmental Institute, 2014).

10  In the AC context, black carbon was from the onset recognized as playing a unique role in the Arctic and should therefore be studied "as a frontier area of science" (Task force on SLCF, progress report 2011). However, the task force carefully explained that such a focus is not a judgment of this pollutant being of greater importance than methane or any other climate forcers.

11  This interview has been conducted by the Arctic Council and published on the Arctic Council webpage.

12  Although not questioning the positive effects of emission reductions, Russia, for instance, employed a line of reasoning where black carbon and methane reductions not necessarily followed the same logic concerning benefits: whereas black carbon would be bad for health, and methane for environment, one could not draw conclusions in reverse, saying the same causal impact would be valid no matter which short-lived climate pollutant. Denmark raised similar objections, in the sense of warning that political ambitions and decision-making would anticipate science, being, at worst, contra-productive.

13  Interviewees give an account of a very long waiting period between meeting number two in Stockholm in December 2013, and meeting number three in

Moscow in April 2014, which caused the process to lose momentum. A contributing reason for the long period of time between the meetings was a rescheduling of meeting sites, where the meeting initially was planned to be held at the American Embassy in Helsinki but had to be cancelled due to Crimea, and instead moved to Moscow. Neither Canada or the U.S., nor the permanent participant AIA, were present at the Moscow meeting (TFBCM Two-pager, Moscow 2014).

14 Referral is made to the Helsinki Convention, which is a regional environmental convention from 1974 to 1992 for states in the Baltic sea area, cooperating under the supervision of Helcom (Helsinki Commission).

15 Norway, here, foremost highlighted the high share of methane emissions within the oil and gas sector, where the potential for reductions was embedded within things such as upgraded equipment and changes in operational practices. Engines, compressors, pipelines, storage tanks, wells, and so on are examples of those components listed as relevant mitigation measures (Task force on SLCF, 2013:7).

16 This interview has been conducted by the Arctic Council and published on the Arctic Council webpage.

## References

AC Ministerial Statement by Canada. (2015). *Address by Minister Aglukkaq to Arctic Council Ministerial Meeting on Canada's Chairmanships Achievements.* 10th Ministerial Meeting. Iqaluit, Canada, April 24, 2015. Accessed via oaarchive.arctic-council.org.

Arctic Monitoring and Assessment Programme (AMAP). (2015). *Summary for Policy-Makers: Arctic Climate Issues 2015. Short-Lived Climate Pollutants.* Oslo, Norway.

Arctic Council. (2004). *ACIA – Impacts of a Warming Arctic. Arctic Climate Impact Assessment.* ACIA Overview Report. Cambridge: Cambridge University Press, pp. 1–140.

Arctic Environmental Protection Strategy (AEPS). (1991). *Declaration on the Protection of the Arctic Environment.* June 14, 1991, Rovaniemi, Finland.

Blokov, I. P. (2011). *Brief Overview of Oil Pipeline's Ruptures and Volumes of Oil Spills in Russia.* Report: Greenpeace Russia.

Carbon Limits. (2014). *Assessment of Flare Strategies, Techniques for Reduction of Flaring and Associated Emissions, Emission Factors and Methods for Determination of Emissions from Flaring.* By Pederstad, A., Smith, J. D., Jackson, R., Saunier, S. and Holm, T. Final report, M-312 (English). Norwegian Environment Agency.

Danish Ministry of Energy, Utilities and Climate. (2018). *The Climate Initiative in Denmark.* Accessed on January 9, 2018, from http://en.efkm.dk/climate-and-weather/the-climate-initiative-in-denmark/.

Denmark: National Report on Black Carbon and Methane. (2015). *National Report by Denmark, September 2015. Enhanced Black Carbon and Methane Emissions Reductions – An Arctic Council Framework for Action.* National Submission to the Black Carbon and Methane Expert Group, the Arctic Council.

De Souza, M. (2013). *Stephen Harper's Environment Minister Casts Doubt on Climate change.* Canada.com. October 8, 2013.

*Enhanced Black Carbon and Methane Emissions Reductions: An Arctic Council Framework for Action.* (2015). Annex 4: Iqaluit 2015 SAO Report to Ministers. Tromsø, Norway: Arctic Council Secretariat.

Exner-Pirot, H. (2016). *The Arctic Factor: Can Regional Cooperation Thaw Relations between Canada and Russia.* Interview with Huebert, R. and Exner-Pirot, H. In *Open Canada,* February 17, 2016.

*Fairbanks Declaration.* (2017). *Declaration of the Foreign Ministers of the Arctic States.* The 10th Ministerial meeting of the Arctic Council, held in Fairbanks, Alaska, May 10–11, 2017.

Geertz, C. (1980). *Negara. The Theatre State in Nineteenth-Century Bali.* Princeton: Princeton University Press.

Government of Norway. (2015). "Norway Support Efforts to Eliminate Gas Flaring by 2030," News, April 17, 2015. Accessed from www.regjeringen.no/en/aktuelt/eliminate_flaring/id2407055/.

Greenpeace. (2017). *Russian Oil Disaster. The Ongoing Arctic Oil Spill Crisis.* Accessed on June 12, 2017, from www.greenpeace.org/international/en/campaigns/climate-change/arctic-impacts/The-dangers-of-Arctic-oil/Black-ice – Russian-oil-spill-disaster/.

Habermas, J. (1984). *The Theory of Communicative Action. Reason and the Rationalizaton of Society,* Vol. 1. Cambridge: Polity Press.

Hansen, J. (2016). *Danish Agriculture Can Reduce Greenhouse Gases.* Danish Centre for Agriculture – DCA. December 5, 2016, accessed on December 28, 2017, from http://dca.au.dk/en/current-news/news/show/artikel/klimakravene-til-dansk-landbrug-kan-realiseres/.

Hecht, A. D. and Tirpak, D. (1995). "Framework Agreement on Climate Change: A Scientific and Policy History," *Climatic Change,* Vol. 29:4, pp. 371–402.

Interviews. (2014–2015). Interviews conducted between December 2014 and December 2015 with state representatives and other representatives active in Arctic Council cooperation and negotiation.

*Iqaluit Declaration.* (1998). The First Ministerial Meeting of the Arctic Council, held in Iqaluit, Canada, September 17–18, 1998.

Jacovella, F., and Kahn, J. (2015). *TFBCM Co-Chairs Discuss Progress.* Interview published on the Arctic Council webpage January 12, 2015. Accessed from https://arctic-council.org/index.php/en/component/content/article?id=275:tfbcm-co-chairs-discuss-progress.

Kahn, J. (2014). *Arktisk överenskommelse i Tromsö.* Ministry of Environment Sweden – Blog. November 19, 2014.

Kerry, J. (2015). *Remarks at the Presentation of the U.S. Chairmanship Program at the Arctic Council Ministerial.* 9th Ministerial Meeting. Iqaluit, Canada. April 24, 2015.

Ministry of Environment Sweden. (2013). *Chair's Conclusions from the Arctic Environment Ministers Meeting: Arctic Change – Global Effects.* Jukkasjärvi, Sweden, February 5–6, 2013.

Nilsson, A. E. (2007). "A Changing Arctic Climate. Science and Policy in the Arctic Climate Impact Assessment." Dissertation. Linköping Studies in Art and Science No 386. Department of Water and Environmental Studies, Linköping University. Linköping: UniTryck.

Nord, D. C. (2016). *The Arctic Council: Governance within the Far North.* New York: Routledge.

Norway: National Report on Black Carbon and Methane. (2015). *National Report by Norway, September 2015. Enhanced Black Carbon and Methane Emissions*

*Reductions – An Arctic Council Framework for Action.* September 1, 2015. National submission to the Black Carbon and Methane Expert Group, the Arctic Council.

Pedersen, T. (2012). "Debates over the Role of the Arctic Council," *Ocean Development & International Law,* Vol. 43:2, pp. 146–156.

Rosenthal, E. (2015). *Strong U.S. Leadership Can Deliver on Black Carbon Reductions in the Arctic.* Earthjustice – Blog, April 23, 2015.

SAO Meeting Minutes. (2011). Arctic Council Senior Arctic Officials Meeting Copenhagen, Denmark. March 16–17, 2011. Final report, pp. 1–13. (All meeting minutes accessed via oaarchive.arctic-council.org)

———— (2008). Svolvær, Norway. April 23–24, 2008. Final report, pp. 1–17.

———— (2008). Kautokeino, Norway. November 19–20, 2008. Final report, pp.1–16.

———— (2002). Inari, Finland. October 7–8, 2002. Minutes, pp. 1–15.

———— (2001). Espoo, Finland. November 6–7, 2001. Minutes, pp. 1–24.

SAO Report to Ministers. (2013). *Report of Senior Arctic Officials to the Arctic Council Ministers.* Summary report. Kiruna, Sweden, May 15, 2013.

Shapovalova, D. (2016). "The Effectiveness of the Regulatory Regime for Black Carbon Mitigation in the Arctic," *Arctic Review on Law and Politics,* Vol. 7:2, pp. 136–151.

Stickman, M. (2015). Statement by Chief Michael Stickman, International Chair of the Arctic Athabaskan Council, at the Arctic Council's 9th Ministerial Meeting in Iqaluit, Canada, April 24–25, 2015. Available via arctic-council.org.

Stockholm Environmental Institute. (2014). *The Multiple Air Quality and Climate Benefits of Tackling Shortly-Lived Climate Pollutants.* SEI factsheet. Accessed from www.york.ac.uk/media/sei/documents/factsheets2014/SEI%20 Factsheet%20-%20slcfs.pdf.

Stone, D. P. (2015). *The Changing Arctic Environment. The Arctic Messenger.* New York: Cambridge University Press.

Sweden: National Report on Black Carbon and Methane. (2015). *National Report by Sweden, September 2015. Enhanced Black Carbon and Methane Emissions Reductions – An Arctic Council Framework for Action.* National submission to the Black Carbon and Methane Expert Group, the Arctic Council.

Task Force on SLCF (Short-Lived Climate Forcers). (2013). *Recommendations to Reduce Black Carbon and Methane Emission to Slow Arctic Climate change.* Summary Report.

———— (2011). *An Assessment of Emissions and Mitigation Options for Black Caron for the Arctic Council.* Technical Report.

TFBCM Two-pager (2013–2014). *Summary Report.* Task force for Action on Black Carbon and Methane. Meetings one to six, September 2013–November 2014. Accessed from the Arctic Council Open Access Repository: https://oaarchive. arctic-council.org/handle/11374/818.

Warrick, J. and Eilperin, J. (2011). "Arctic Council to Address Role of Soot in Global Warming." *The Washington Post,* May 11, 2011.

World Wildlife Fund (WWF) (2014). *Modeling Oil Spills in the Beaufort, Bering and Barents Seas.* WWF Arctic factsheet.

# Conclusion

## The Studying of Learning through Roles – Findings and Suggestions

### Introduction

This book has had an interest in role-playing, environmental politics, and not least learning in IR. The task for this last chapter is to provide for some understanding of their correlation, drawing on the findings of previous chapters. In so doing, it discusses learning in two different ways; in a conceptual way, placing role theory and the social process of learning in the center of attention; and in an empirical way, connecting learning to the progression (or not) of environmental issues in the Arctic.

This conclusive chapter is set to provide for an understanding of how – and to some extent why – change occurs. The first part of the chapter is theoretically oriented, and develops upon the role as being, to some extent, a reflection of the social context. Having the role concept as a point of departure and drawing on the negotiations of oil spill prevention and reductions of short-lived climate pollutants, it confirms the learning model previously presented in Chapter 2 (Figure 2.3) and suggests *the role to be a flexible construction, informed by situated understanding*. The flexibility is the result of an imbedded wish of role conceptions to stay preserved, while adapting to a context that may be undergoing change. The second part of the chapter is empirically oriented in its ambition to – tentatively – discuss learning in relation to fossil energy interests and climate change mitigation. In relation to the Arctic and Arctic state behavior, some restrained progression could be witnessed, where learning on environmental values *and* strategic adaption both follows the trajectory of a context in change.

### The Correlation between Roles and Learning: Findings and Suggestions

Would learning always be the result of an active search? In Chapter 2, Martha Finnemore was referenced as saying, "states do not always know what they want" (1996:128). This is a theoretical assumption that presupposes that a state's want is – sometimes – intangible. At the same time, social interaction quickly adds information, structures, and expectations in such a way that the social context moves a state's 'want' toward tangibility. Hence, a learning of environmental norms is not only, at times perhaps though not at all occasions, a matter of the specific environmental issue but of social interaction and those state positions being upheld therein. Indeed, when it comes to learning in general and learning of environmental norms specifically, the importance of situated understanding moves learning beyond the environmental field per se into the social field. One of the participants interviewed in this book, with experience of acting as task force leader, illustrates this by pointing towards understanding as being a (partly) result of social interaction, and describes how the distance between actors *de*creased as understanding *in*creased:

> [after some negotiation], there is improved understanding. Not that one is adjusting to each other, that one is more faithful to the process than the country…Sometimes one may think that is the case, but no. No, it is about understanding, improved understanding of each other's arguments.

The interviewee separates the wish for staying loyal to its own country from an improved understanding derived from social interaction. The notion of negotiating states staying loyal to the norm on sovereignty is benchmark within the interactive logic where it is "appropriate for actors to pursue their self-interest" (Müller, 2004:416). However, this ad hoc relationship between appropriateness and self-interest has also been concurred to change should it collide with a valid norm that prescribes behavior differently (ibid.). The empirical investigations of Chapters 4, 5, and 6, have found one such collision to be precipitated by a strong – and by the Arctic states perceived valid – norm on cooperation. Therefore, and to go back to the quote by the co-chair, to remain loyal to its own country cannot be separated or be differentiated from an improved understanding that has to do with social re-evaluations. Theoretically, this means that a question such as how state X (and its role) will be affected if doing A, B, or C is answered by stretching beyond ego to also take alter expectations into account: in a process like a negotiation, alter expectations are by necessity always involved should ego – and its wants – be truly elucidated. In a more practical sense, to the Arctic states, this means that they cannot stay loyal to their own countries without simultaneously thinking of – and staying attuned to – Arctic cooperation.

This book has suggested roles to have a bearing on the way Arctic environmental cooperation is developed, and ultimately, on learning. States are filtering norms and expectations (i.e., prescriptions on behavior) against their role conceptions, in search for what would be considered an appropriate – and to some degree levelled – behavior. In Chapter 2, a triangle on state learning was introduced, based upon the correlation between roles and learning (see Figure 2.3). Drawing on this, three conclusions will be discussed in the following. First, roles are stable but flexible. Secondly, the flexible dimension of the roles is activated through actors' understanding of their social context. And thirdly, states cannot learn in a speedier and more thorough manner than the role allows for flexing. Taken together, this tripartite suggestion sheds light on how any learning that is occurring is a result of ego and alter expectations, where role flexibility becomes a means for expectations to handle tensions between, on the one hand, norms and structures pushing for change and, on the other hand, actors' preferences for continuity.

### Roles Are Stable, but Flexible

The finding that roles are stable is in line with how Stephen G. Walker has described roles to be "repertories of behavior" (1992, cited in Thies, 2010:6336), i.e., something that includes continuity, performing the role as rehearsed. The role mapping of the Arctic states' role performance in the Arctic Council and at SAOs meetings (Chapter 4) also confirms a stability in these actors' behavioral repertoires. Changes were not so large as it would be warranted to talk about states adopting a new role in the AC interaction, but rather, as deploying a behavioral pattern visible over time. The finding that roles are stable is also captured by Sebastian Harnisch through the somewhat parallel and yet intertwined process of norm development and role stability, where he describes roles as social positions that are "constituted by ego- and alter- expectation regarding the *purpose* [emphasis added] of an actor in an organized group" (Harnisch, 2015:5). Consequently, as context changes, actors may have to change their behavior should the purpose – its role characteristics expressed as social position – be kept intact. An even more central finding would thus be that although roles are stable, *they are flexible*.

The Arctic states seem to have adapted their role performance to a context that is under constant development, in turn, spurring the development of new role strategies of those interacting. Different strategies align differently with the role, displaying the latter's flexibility. For instance, Norway's leadership has become more restless and less listening as its know-how has been confirmed and its wish to see results has increased – both environmentally and energy extraction wise; Russia's (sovereignty-based) informative component has been toned down as Arctic cooperation has moved into a period of more political insecurity, and Finland is keen on taking responsibility but, contrary to the – in comparison quite unpolitical – beginning, no more as a leader, exempting periods of chairmanships. In that way, roles are stable, but the potential role conflicts that may emerge as social interaction proceeds require of states to develop new role strategies to make the most out of the role, given the social setting. That roles are flexible does not preclude the significance of having knowledge about them, since even the most flexible role retains a core of stability and continuity in perceptions and behavior. The knowing of roles, therefore, improves understanding of prospective environmental cooperation.

The role-set that dresses a state is big to wear and needs to be made fit the occasion. The most telling example of the role repertoire as encompassing multiple dimensions becomes visible when comparing the Arctic state's role patterns in the AC, with their role behavior outside of the Arctic. Internationally, the U.S. and Russia are described as suffering from 'irreconcilable differences,' something that renders cooperation and mutual understanding difficult (Carlsson and Winnerstig, 2016). Likewise, the EU and other Western states condemn Russia which not only acts in a militarily aggressive manner and is disrespectful of territorial borders or airspace, but is also viewed as politically dishonest by infiltrating national election campaigns (European Council, 2018; U.S. Office of Director of National Intelligence, 2017). The Nordic states, furthermore, are generally all small states but gain in size through normative entrepreneurship (Ingebritsen, 2002; Magnúsdottir and Thorhallsson, 2011). These are roles performed outside of the Arctic Council, but *inside* the Arctic Council, small states also become 'big,' big become 'small,' and 'enemies' become allies. That the Arctic is unique, in terms of bringing forward cooperativeness among actors that are not always acting cooperatively, has since before been acknowledged and referred to the AC's attempt to cultivate common interests (Nicol and Heininen, 2014; Heininen, 2015; Albert and Vasilache, 2017). Therefore, Arctic kinship, be it materially or normatively informed, seems to shrink the role-set considerably to make it fit a region of mutual interests and concerns.

No matter what, the fact that the Arctic states share an interest of cooperation does not spare roles from conflicts and tensions. The negotiation process of TFOPP, described in Chapter 5, revealed how states that were materially and territorially big acted as leaders, raised their voices, and had their role performances to reflect upon them having 'claims' in the Arctic, territorial or resource wise. To perform the role of, for instance, the Responsive informant, Protector, or Know-how leader, did in this negotiation setting pertain to sovereignty and to issues of oil extraction and maritime transport. The 'meeting dramaturgy' set the frames for what would be possible to say and by whom (see Campbell et al., 2014) in such a way that the rules of the game – role prescriptions – left non-littoral states to play a less visible role: to be a Reserved team player like Finland, or Teacher on demand like Sweden, signified keeping a low profile, to provide guidance and expertise when requested, but without going against the prescription on paying respect to (littoral states') sovereignty. The negotiation process of TFBCM described in Chapter 6 prescribed a somewhat different behavior. Here, there was a firmer demand for action, which, for instance, made it possible for Sweden 'to grow' and act like a progressive teacher. Norway and the U.S., furthermore, tilted their leadership to fit this particular issue of short-lived climate pollutants, where they provided for know-how and scientific leadership in a confident manner; Canada protected its inhabitants against an unfair distribution of commitments; and Russia had little to inform on regarding environmental progress and thus remained silent. Whereas Canada and Russia had voiced political disparity in the TFOPP, the topic discussed in the TFBCM had them consigned into something close to understanding.

To link role flexibility to learning, the argument is presented on the need for a micro-perspective to catch cognitive processes. Because, whereas behavior would be enough to study as *role performance*, such behavioral observations are not sufficient to provide for an understanding of learning and how this behavior came to represent that state; the interesting thing is not to depart from what the eye can see, but which beliefs are held (Mead, 1925; 1934. A role incoherence could be derived – and made visible – from what in this book has been referred to as *I*, or 'ego.' Norway, for instance, has two *Is*, one pertaining to being somewhat of an Arctic environmental rescuer, and one pertaining to be a knowledgeable energy and maritime actor. When the issue of oil spill prevention was negotiated in TFOPP, Norway faced difficulties when its 'energy *I*' was targeted by others' expectations of being an environmental actor, i.e., someone who should act in the name of environmental protection. Norway yielded in, even though the TFOPP context more than anything prescribed behavior on oil spill know-how. But one did so because both of these *Is* – environmental protector and resource extractor – are part of the Norwegian Self, and here collided. Also, states like Canada, Denmark, and Russia did all dedicate their efforts at trying to strike the correct balance between being a good cooperator while still protecting sovereignty and ensuring security for the inhabitants. Role flexibility, therefore, is a means of social response and, as such, a result of social learning.

### *Understanding – The Mechanism Activating the Flexibility in Roles*

Actors that argue are exposed to role prescriptions and others' expectations in a direct way, getting in touch with the alter role part via language and communication. Arguing could be both a promise and a curse, depending on the amount of 'commonness' between actors; sometimes arguing brings actors closer to one another, and sometimes astray. Environmental cooperation in the AC is aligned through two different understandings: one connected to 'cooperation' and one to 'environment.' The latter has more room for interpretation, and therefore causes more arguing, due to containing less stringent prescriptions. The micro-level analyses of this book have revealed states to share a common understanding on the need for cooperation, whereas they have turned to arguing in order to settle the discourse on the prefix 'environment' in environmental cooperation.

#### *A Common Understanding Oriented at Cooperation*

The negotiations of oil spill prevention and reduction of short-lived climate pollutants both illustrated how consensus-reaching was guiding the Arctic relations. The presence of such guidance was similar to interview accounts such as: "everybody was trying to find solutions rather than confrontations;" "it pushed Arctic relations, the understanding of us having something peaceful and unique"; and "with four years of work behind us, we were many who thought it was time to progress." Even when the incident in Crimea deteriorated the relations, when political boycotting of AC meetings occurred and the negotiation climate went worsened, states stayed committed to consensus-reaching. "Gradually, it became more important to get an agreement, than what, de facto, was in it," a TFOPP participant explained. For sovereign states, with strong interests in the Arctic, strategic action should not be excluded as the driver for cooperation. Acknowledging that, however, does not mean to preclude any other interactive logics to guide cooperation and negotiations in the AC, and it is suggested that negotiations, to some extent, were indeed oriented toward a common understanding on the norms of cooperation.

In Chapter 2, the likelihood of communicative action was explained to linger on the following: the thicker the institutional environment and the more moral entrepreneurs that are present, the more salient communicative action would be, and vice versa (Müller, 2011). In the Arctic Council, as outside, several states are acting like moral entrepreneurs. Other Arctic states, however, are not. In addition, although some policy-making has been accounted for, the institutional structure of the Arctic Council cannot be defined as thick. Yet, the empirical investigations of Chapters 4, 5, and 6 have revealed eight states – or eight *roles* – to care for cooperation. This pertains also to those states who normally and actively do not take on roles as norm entrepreneurs. How could that be? A suggested reason would be that it is the *implicit* rules that are thick to their character. "It is an equal process," a Scandinavian delegate explained, "everyone is acting concessive for the sake of the greater task." The greater task would be cooperation, since Arctic state interests all require this to be materialized. "To reach an agreement in the AC is an easier forum, more common interests between these states," another delegate concluded, comparing the AC to other international forums. On the Ministerial level, statement after statement, year after year, confirms the good relations between the Arctic states, and how vital and essential Arctic cooperation would be for them all. The Norwegian Foreign Minister Ine Eriksen Søreide more recently concluded that it would be no coincidence that the Arctic region is characterized by peace, stability, and international cooperation, but rather, would be "a result of political choices. Choices made by every country around this table" (AC Ministerial statement by Norway, 2019). It could be determined that Arctic cooperation is of profound national interest to the eight Arctic states.

Although common understanding has been oriented toward benefits of having cooperation guide relations, cooperation in the Arctic Council has not reached a stage where it is completely self-explanatory. In the TFBCM, delegates felt doubts all the way through regarding other states' true level of commitment. And in the TFOPP, furthermore, a distance was created to specific actors (i.e., Russia). The understanding of 'rogues' being part of the process has therefore been constraining cooperation, questioning the sincerity of the actors involved. Such a lack of trust has been expressed through the continuous suspension of dinner meetings prior to Ministerial meetings, correlating in time with the annexation of Crimea. "These meetings were very useful for building trust and mutual understanding," the Russian Foreign Minister Lavrov regrettably explained (2017). This observation has also been noted by others, where, especially in the early days of the Arctic Council, state officials and others involved considered the structure of the AC to be informal and relaxed, where meetings and interaction happened through the "small" and more personal dimensions of international politics (see Spence, 2017). Thus, whereas all states play the role of being cooperators, this is a role-casting that – at least presently – causes some unsureness.

*Settling the Discourse on 'Environment'*

Elsewhere in this book it has been discussed how the Arctic future would be seen as without predestination; it is open-ended, allowing for actors to fight over the outcome (Wormbs and Sörlin, 2013). Thus, will the Arctic future be governed through political convictions or by scientific findings, or perhaps with both combined? Indeed, the environment, and thereto environmental threats, is added values through political understandings on 'what goes,' which is to be determined through a trisected validation process in relation to the objective, social, and subjective world (Habermas, 1984).

Today, oil spills and short-lived climate pollutants illustrate two issues where the discourse concerning 'what goes' is not yet settled but is dependent on arguing for a consensus-reaching. These two negotiation processes both recognized their respective environmental problem as a result of *objective* knowledge and agreed to *think* of it as a threat: "the common nominator between us was the will to do something about it [short-lived climate pollutants]" or, "the best response to [oil] pollution is to never have it happen in the first place," delegates explained. Both negotiations also revealed communication to be faulty in a *subjective* sense, where understandings of the other state members did not support enough and justify a common approach. In the TFBCM, a lack of perceived sincerity amongst participants caused concerns about others' emission reduction commitment: would they take measures, or would they try to get a free-ride, or even depart? And in the TFOPP, a lack of sincerity made actors doubt the reasons behind cooperative ambitions. Participants questioned the frenzy with which Russia pushed for a legal agreement, and there was hesitation toward sharing information across the border: "if one is sort of suspicious regarding what the information will be used for, then obviously one will be very careful in what to agree to," a delegate explained.

Arguing, therefore, mostly differed in the two task forces in relation to the *social* world, where understanding is getting affected by each state's economic and political reality. The big difference between TFOPP and

TFBCM lies in the former's applied strategy of *avoiding* arguing to the extent possible, to lessen the risk of having states pushed further away from the overarching goal of cooperation and of staying united. Even though oil spill *prevention* was the objective, the negotiation process did not challenge the prevalence of Arctic energy extraction or other economic activities by having such activities discussed from a legitimacy perspective. Plausibly, the TFOPP expected negotiations to become too tension-filled, should participants argue on issues infringing upon sovereignty. This was illustrated by one participant saying: "it is important to act now, while there are only economic interests to constrain and not yet any real economic activities." Consequently, the negotiation process applied a strategy circumventing the platform from where resistance (i.e., ego) could pinion a consensus.

In the TFBCM, arguing went deeper into states' social worlds, allowing for greater space for the ego to partake in discussions. The presence of ego was noticed in the raised objections being made, similar to, 'territories are too big to overlook and control for emissions,' 'specific states already contributed more than others,' and 'flaring is too linked to the energy sector for us to decide on here.' The arguing process revealed a resistance to validate (all) knowledge as true in the social world. However, the TFBCM had high ambitions, and it was willing to introduce the social reality of each state – allowing for more resistance but also revealing more opportunities to change. It should not be seen paradoxically that the more ambitious results the states are after, the more conflicting the negotiation process could be. But if succeeding, the greater are the gains. This finding suggests arguing in the longer run to be a possible game changer; the more actors argue, the more likely it is that change will come about – as long as actors do not deviate but stay put in the negotiation process.

### *Language and Communication – Key Components in Learning Processes*

When interviewees were talking about their respective task force process as involving learning, they mostly had in mind a learning pertaining to the subjective field, where they got to learn more of each other. Words, in such a process, could both ease and freeze. Clearly, linguistic details were explained to be big; to some, the obsession with words were 'silly' attempts for power, to others it was dead serious. Small words also had a bearing on how the world is socially understood, which one North American delegate showed in the following way: "sometimes people get stuck with little words. A word means something to one person, but it means something different to someone else. ...It is a word, but words mean a lot." Words, in this sense, can turn into a prolongation of performance, a protest in the small when the 'big' is calling for a consensus.

Language – how words are used – is also suggested to have the effect of creating more harmony by having misunderstandings erased. Interviewees talked about a greater understanding that arose once they had figured out the true essence in other delegates' choice of words, been acquainted with their body language, and learned to use correct phrases similar to others. "It becomes more constructive at the end, after one has gotten to know each other. In the beginning – then, it is difficult," one Danish delegate explained. Subjective understandings of others' sincerity, so it seems, is therefore not disconnected from how languages – words – are used. It is also through communication that actors became aware of what 'a good cooperator' is meant to be. In the words of another Scandinavian delegate, who acted as an adviser:

> One adapts to the allocated role. Not like being told what to think, but one yet [adapts] at informal meetings. I give my view on something, and get feedback – 'this is good, we should bring that up, let us present it like this'…Not that I had to think differently, that has not been the case. I have been able to think or say whatever I want, but it has been acted upon differently.

This quote points toward recognition as a socializing tool. The adviser is free to think whatever he or she wants, but while thinking freely he or she still learns which opinions are appropriate (as in doable), given the context and given the role as upholding a certain social position. Following the human need for self-esteem, the thing the actor is recognized for – if attached positive values – would likely be a thing that the actor then also would learn to repeat. In that sense, communication does for ego and alter learning what values do for environmental norms: they bring the substance about.

### Summing Up: The Linkage between Understanding and Role Flexibility

The prospects for validating knowledge as true in all three 'worlds' spoken of by Habermas (1984) are suggested to be shaped as a funnel: to accept norms as objectively true comes first, being the – in the context – easy part. To arrive at a claim being validated true in the social world is less present and trickier to achieve. Lastly, the validation of knowledge as true in the subjective world seems to quite a high degree be hindered by a lack of sincerity, throwing a shadow over cooperation at large.

In relation to role flexibility, arguing and understanding is suggested to activate different dimensions of the role. The more accessible (common) understanding provided for in the objective world does correlate with role prescriptions, i.e., actors are made aware of expectations and the rules of the game and do, rather easily, confirm these as being true. In relation to the social world, the incentives for arguing increase. Here role conceptions are partaking and could either try to push for continuity or for change. In Chapter 4, all the Arctic states were revealed to have a role conception of them being good cooperators. But states were not equally clear on what another recurrent theme within role conceptions actually meant – that of being a protector of the Arctic environment; thus, the environmental discourse was less established, less common, than the cooperative ditto. This signifies role-flexing to depart from what is an appropriate behavior for a *cooperator*, pushing states to firstly reflect on what type of cooperative behavior is expected, and *then* adding environmental norms as appropriate. It is suggested that conceptions of being a good cooperator is the constant, where norms on environmental protection are merged into such good cooperative behavior. But, if set free, environmental norms could also attempt to challenge the understandings of what constitute such an actor, pushing environmental cooperation forward by bringing in new values for actors to consider.

The TFOPP revealed a process centered on 'cooperation,' where, for instance, Finland and Sweden took a step back on environmental protection and progressive calls on stricter definitions of prevention; 'Why should we go around and act like wiseacres in an area we are not industrially operating within,' they reasoned, and gave priority to the act of cooperating over oil spill consequences. In the TFBCM, the same states had their roles flexed back to the environment and requested change through progressive environmental commitments. In contrast, several coastal states in the TFBCM agreed to emission reduction measures that others did not really expect of them, or at least, did not take for granted.

When it comes to international cooperation and social interaction, states, so it seems, behave like rule-followers. Yet, this is not what the states always expect of others to do. Indeed, in the subjective world, states, until proven wrong, tend to perceive others as being far less obedient (i.e., not as rule-followers). This indicates that experience encompassing relations – long-during and trustworthy behavior – should be a beneficial component within environmental cooperation, at least if it should be progressive and where actors could learn of environmental norm in such a way that environmental discourses are settled to include real commitment on protection.

### *Learning of Environmental Norms: A Time-Consuming Process*

Chapters 5 and 6 revealed how role prescriptions and alter expectations could activate the flexibility within roles. To flex the role is an expression of all states and actors wanting to have their role – social position – preserved in accordance with role (ego) conceptions. As such, it has been suggested that role-flexing itself would be an expression for ego and alter learning, and thus ultimately, social learning. As we move away from the process of learning toward learning *as a result,* the adaptive learning correlates with a change in role performance whereas the normative learning – which targets beliefs – correlates with a change in role conception. Accordingly, it could be concluded that the U.S. leadership on black carbon was not the result of a normative learning targeting the belief system, since once a new presidential administration took office, leadership was pulled back. Therefore, the U.S. leadership on black carbon for sure could pass as role performance, but it could not make it the whole way through to role conceptions and become an integral part of U.S. innate leadership role. Following Cantir and Kaarbo (2016), a foreign policy behavior where the U.S. is acting like a climate leader is not a behavior being anchored and supported across political axis but affected by a still ongoing horizontal role contestation between national political elites.

Learning, just like understanding and just like roles, is intertwined in social processes lacking clear beginnings and clear ends. Whereas different learning types can be crystalized in theory, they are generally more difficult to differentiate empirically – really, what would be a result of normative learning and what is not? As the Arctic states have cooperated in the AC for more than two decades, it has moved from (only) normative decision-shaping and political awareness-raising, to more political decision-making and increased Arctic governance through three binding agreements. In addition, Arctic states have also been introduced to, aware of, and socialized to different environmental issues such as Arctic warming, oil spill prevention, and resilience, to mention a few, while the AC simultaneously has expanded its agenda to include issues beyond the environmental field, like research and maritime issues. As pointed out by a former SAO of Finland, Hannu Halinen, "the growing normative role of the Council is a fact" SAO meeting Portland, 2016). Thus, that a normative learning has occurred also amongst the Arctic states should not be ruled out. One of the PPs, the Saami Council, describes how values such as environmental protection and sustainable development have been a constant throughout AC's history; "this value foundation has not changed, only more people have come to share those values" (AC Ministerial statement by Saami Council, 2017). But neither could it be ruled out that AC development would be a result of strategy and adaption, given states' desire to remain as 'cooperators' also in times of raising environmental demands and a changed context. Actors may need to adjust their behavior – strategies – in order to make the most out of the situation and in relation to respective (ego) role conception.

On a general level, time seems to be a deal breaker. The amount of time dedicated to a task force process – five to six meetings during one to two years – is too short if substantial normative learning is to occur. Because the thing with normative learning is that it cannot happen quickly should ego conceptions agree; normative learning presupposes time and repetition where the role is allowed to adapt step by step to the extent where 'change' becomes transformed into 'continuity.' Because through such a stepwise approach, roles can change in content – be flexible – without losing sight of their position; their conceptions of who they are in international relations would be kept intact. A hypothetical but illustrative example would be Norway: that this state would ban Arctic oil extraction because those sitting at a negotiation table would demand thereof should be held unlikely. *I* would react violently and impulsively against such a demand. Instead, there needs to be time enough for a strategy to be developed, where the role as a knowledgeable and a great (i.e., big) Arctic know-how leader could be stretched enough to be maintained, despite one source to such role input has gone missing.

Empirically, to be able to observe learning on environmental norms – whether adaptive or normative – takes a change, a before and after, in international relations: should, for instance, the price of coal be drastically dumped, attracting a lot of new investments while investments in renewable energy as a consequence would stagnate or detract, it seems like international learning on climate change after all would not have been that significant. Industrial talk and action on climate change would then not have its base in beliefs and strong convictions, but rather, would be a result of strategic adaption to a governance structure that prescribes considerations taken thereof.

Yet, an empirical distinction may not be that important, after all. The environment does not care if cooperation is executed with nature's best in mind, or if it is a strategic decision based upon utility maximization: the only thing cared for by the environment is to arrive at enhanced protection. Also, adaptive learning may serve to confirm environment with values, which eventually even could lead to normative learning as the pressure from prescriptions would increase.

## Learning about the Arctic Environment: Restrained Progression

Elsewhere in this book, the drifting ship of Nansen, frozen into the sea ice in an attempt to reach the North Pole, was used as a metaphor for movement, possibly learning. All social interaction unavoidably leads somewhere, but should it be such directional movement that Nansen hoped for it takes a movement driven by learning, i.e., a movement striving toward a specific, and ultimately a common, goal.

In the Arctic Council, cooperation is not adrift. "We have been able to sail together," Finland concluded its chairmanship: "perhaps we sometimes have had windy conditions – but that is what keeps the ships moving" (AC Ministerial statement by Finland, 2019). As a matter of fact, 'windy conditions,' or arguing, is what pushes Arctic cooperation forward. Under the auspices of the Arctic Council the fragile Arctic environment has been paid attention in ways previously not thought of. Not least the indigenous associations – the PPs – have added information and knowledge from a local perspective, and their influx to actions on, for instance, Arctic warming and sustainable development has been consequent from the very beginning. To view and incorporate Traditional Local Knowledge (TLK) has increased in frequency during the AC's development, and constitutes, alongside science, nowadays also Arctic expert knowledge.[1] If this trend continues, the influence from the indigenous communities should be expected to grow in the future.

As stated elsewhere, the Arctic states have learned how to play the 'Arctic Council game' in a way that presents themselves as being cooperators. Through such a maneuver, Arctic interests – national and/or common – are rendered easier to obtain. The Arctic states have come to be aware of the importance of reaching consensus, albeit the symbolic action of reaching consensus comes prior to content; orchestrated performances on unity and control are paramount not only for the well-being of Arctic relations but also for the way external actors relate to the Arctic. From a strictly environmental perspective though the cooperative course has not been as straightforward as wished for. The AC is a consensus-based organization, where voluntary decisions and commitments govern. It works in a normative manner by enhancing knowledge on environmental issues, but concrete decision-making and action-taking is limited in scope. In that sense, an international learning – as in shared learning (see Knopf, 2003) – about how to approach and protect the Arctic environment reveals restrained progression.

### Environmental Protection – Belief and Strategy Combined

According to Falkner (2012), there is an ongoing greening process in international society. For actors with ambitions of being recognized as legitimate, environmental norms cannot be completely disregarded. Such socialization could reasonably be believed to apply also to the Arctic states, which care for being perceived as good cooperative partners in an environmentally sensitive region. The international contestation of global governance norms has created a specific version of Arctic 'greening,' where states are obliged to pay attention to oil spills and Arctic warming. WGs, AC Ministerial statements, and declarations have in an iterated manner pushed for increased knowledge and understanding of these issues more or less from the AC onset.

Over the years, the tone on Arctic states' responsibility to *do* something – not least in relation to Arctic warming – has become considerably sharpened. In a joint statement, PPs have, for instance, voiced their "sincere concerns" over a lack of concrete AC actions regarding emissions of carbon dioxide (Letter from PPs, 2014). But the tone has gone sharpened also by the Arctic states, where an argumentation on the Arctic as valuable – not as a resource but as a common good – has gained stronger ground. A good illustration thereof is found in the AC Ministerial statement where Sweden's Minister of Foreign Affairs rhetorically asks what the planet would say if she could have a seat at the Ministerial table:

> Perhaps our planet would say: I have been your best friend since the Industrial revolution. I have done all I can to dampen and absorb. I have tried to keep Greenland and the permafrost in Siberia intact. I have sent you no invoices. But this is about to change
> M. Wallström, Minister of Foreign Affairs, 2017.

When climate change was introduced to the Arctic Council in the late 1990s, most states did not feel compelled to accept immediately. Twenty years later, it has been rooted in the beliefs of the Arctic states, such that a rhetoric describes nature as superior to the Ministerial table, as someone – and justly, so the quote says – criticizes past and current exploitation. The words signify a general greening trend, an acceptance of norms as representing conventional knowledge which is also being illustrated at the last Ministerial meeting (2019); for instance, Sweden spoke about the passing of a tipping-point of awareness, Canada about new terrifying scientific findings concerning climate change in the Canadian north, and Russia acknowledged the importance of minimizing man-made environmental challenges. PPs, furthermore, sharpened the tone of climate prescriptions further by drawing attention to indigenous communities being in severe trouble due to climate change, that climate change is a man-made threat politicians cannot keep hiding from, and where blinders once and for all should be taken off to urgently make real political commitments (AC Ministerial statement by AAC; ICC; Saami Council, 2019).

This greening process contains, and is enabled by, an acceptance of new norms. A substantial part of this acceptance could merely be about strategy; energy actors are aware of the fact that whether or not Arctic energy extraction will be seen as legitimate is conditional upon their abilities to convince others of them still acting in an environmentally responsible manner. To succeed takes a certain amount of behavioral or rhetorical adaption. However, there is also some support of this greening process being controlled by norms that stretch deeper into state conceptions regarding right or wrong, embodying changed beliefs regarding which values to attach to nature. Significative, when the USGS in 2008 estimated the Arctic as a global resource winner, this could have altered Arctic interaction toward full focus on resource extraction. Instead, environmental cooperation stood firm. It even became intensified the same year, by introducing short-lived climate pollutants which have led to concrete emission reduction targets – and actions. As of 2016, the Arctic states reported, except Russia which so far is lacking projections of black carbon emissions, a total and collective reduction of black carbon by 16 percent compared to 2013 levels. By 2025, the states should collectively achieve a 25–33 percent decrease, and the states seem to be on track although more things need to be done[2] (Expert Group on Black Carbon and Methane, 2019). Yet, the most telling example of the involvement of normative learning in AC cooperation is perhaps the amount of adaptive learning in the Arctic. Had it not been for states and other actors acknowledging new environmental prescriptions and cues concerning appropriate behavior, it would not have been necessary to adapt behavioral strategies in the first place.

The growth of environmental norms is, amongst others, visible in the AC's stewardship role (Pedersen, 2012; Wilson, 2016), which has gained increased attention. All Arctic area is not territorially exclusive: not most of the surface of the Arctic Ocean; not all of the polar areas yet covered in ice, and not the atmosphere. In one way, the Arctic is a regional common;

in another way, it is a global common. State delegates interviewed for this book have given account of it being important to show the global community an Arctic willingness to take on a specific environmental responsibility in the Arctic. This could be interpreted as a will to 'lead by example' and is also applicable to the adoption of the convention banning unregulated fishing in the central Arctic Ocean (2015), and the establishment of Arctic Offshore Regulators Forum (2015). Regional leadership and approaching the Arctic as a common comes at the expense of sovereignty is getting slightly weakened, but at the same time it gains by sending global signals of the Arctic states as legitimate actors who act out on Arctic responsibility. Not the least does it send signals of the Arctic states as being the ones who on their own set out the rules to govern on who exercises Arctic ownership. In this sense, stewardship could be understood as strategic and adaptive learning to maintain the Arctic as a regional common under Arctic state control.

### *The Future of Arctic Oil and Gas: Time to "Expect the Unexpected"?*

Throughout the world, international greening is tuning in calls for emission reductions and divestments in the fossil fuel sector, in addition to a more sustainable living. For instance, in Sweden and elsewhere, air travel has come to be a symbol of humanity's unsustainable relationship to nature, where the environmental impact of such travelling is measured in square kilometers of melting Arctic sea. In 2019, Greta Thunberg, a young climate activist from Sweden, was awarded Person of the Year by the *Time* magazine; what began as Friday school strikes for the climate grew to be a global climate movement where millions of people joined in. Looking straight at the gathered assemblies of world leaders, Thunberg repeated her message over again: "given what we scientifically know, how dare you not to act?" As has been discussed in this book, to learn about environmental norms by sharing – believing in – their values takes not only (scientific) knowledge but also gained insight. Perhaps it is with this insight in mind that the international society now moves forward, in what furthermore seems to be a speed previously not seen. And if public opinion, truly, is contesting the foreign policy behavior of the respective state, a new and more climate progressive role conception – and role performance – may indeed gain ground.

Actually, given recent development and subsequent expectations of how the market will be affected, the fossil fuel industry has been recommended to scale down future demand scenarios, and instead start to "expect the unexpected" – that fossil energy could rapidly become dated (Grantham Institute, 2017). Although peak oil is debated, the tipping point has been reached where solar and wind production has had volumes big enough to reach lower costs per unity produced, than had oil and gas (World Economic Forum, 2016). As an illustrative example thereof, a company like DONG (Danish Oil and Natural Gas), which for over forty years has had its base in fossil fuels, is now the recognized owner of the world's largest offshore wind farms (Clark, 2017). In relation to the Arctic specifically, the World Bank has announced that in line with the Paris agreement it will no longer finance upstream oil and gas, i.e., exploratory or extraction activities (2017). For similar reasons the French oil major Total has presented a new development strategy excluding operating in the Arctic pack ice; and in Norway, one of the oil industry's closest of allies, the Norwegian Labour Party, has decided to withdraw its support for Arctic offshore drilling in the Lofoten archipelago – a resource development the industry deems necessary to maintain current levels of production (Schober, 2019; Staalesen, 2016).

Yet, obviously, the O&G foresee fossil fuels to represent the dominant energy source well into the future. As of 2019, the world still has not witnessed a downward trend concerning emissions from carbon dioxide, but rather the emissions have been constantly increasing. Fossil fuels currently constitute the source of about 80 percent of all energy consumed globally; the largest source of emissions remains coal, but it is increasingly being replaced as oil and even more so natural gas has boosted on the market. Actually, the demand for oil and natural gas is expected to grow until the year 2040 (IEA, 2017). From this perspective, Arctic energy reserves are likely to remain relevant for a foreseeable future.

Vital for Arctic drilling decisions has been the cost argument. The energy company Shell, after having spent seven billion U.S. dollars, has, for instance, put an indefinite ban on Arctic drilling due to too marginal findings (Macalister, 2015).[3] For some time, the oil price has been low, which viewed in relation to Arctic drilling operations is perceived as driving states to abstain from Arctic activities. Only Norway is currently moving forward in its position, but Norway would also be the state with the lowest extraction costs and most favorable climate concerning drilling conditions. Now, from an environmental protection perspective, to pause drilling as the effect of a low global price on oil is not a stable learning outcome, and intensified activities – more prospecting, more extraction – could therefore be expected should the oil price significantly rise. Indeed, the Arctic O&G industry *is* providing a positive attitude on future drilling. It is a business that refers to 'hard facts' and 'good science,' relying less on environmental norms and more on pragmatism regarding what is possible to achieve, and where investment strategies based on demand are expected to remain for at least another 20–30 years.

This said, it should also be noted that the industry is affected by international greening, by being pushed to adapt to a world and market where new norms are emerging, even governing. The industry does what it can to not be perceived as standing in contrast. To rank the reduction of greenhouse gases as number one priority is common amongst oil and gas companies, for instance, Canada's Oil Sand Innovation Alliance which works to make Canadian energy part of sustainable development (2020), or Equinor, who explains that whereas "some people are still disputing global warming. We're acting on it" (Equinor, 2020). In addition, logos or images are often signaling the caring of a healthy planet. Adaptive strategies are clearly noticed, for instance, through Equinor marketing of its entry into the Barents Sea petroleum exploration as an act of responsibility, given their advanced and safe drilling operations in a time where people globally suffer from energy poverty (ibid, 2019). The Norwegian Oil and Gas Association agrees, describing climate change as one of the most important issues to solve, just as important as it is to increase the energy supply to encompass those two billion people still lacking electricity (ibid, 2017). The oil industry is also investing large amounts in research, for instance, on oil spills prevention, which – in the case of the TFOPP – provided the O&G with the opportunity to let all their protective work be visible for those outside of business.[4]

The industry's interest in oil spill prevention can be referred to as self-interest. That this self-interest is also guided by a greening process is evidenced by a task force such as TFOPP, which is shedding light on commercial activities in the Arctic – like oil extraction and maritime transport – as being conditional upon no accidents to (preferably) occur. That this learning (also) is adaptive and influenced by strategic reasoning was noticed in the way that the task force participants reasoned; should an oil spill occur the whole of the Arctic would be affected – pollution wise but also by the very negative response from the environmental groups – independently of where the spill would occur. The industry itself approaches accidents as bad business – costly to clean up and bad for future investments, and where one accident can shut down the whole of Arctic O&G activities. In addition, Norway reports of the difficulty in attracting personnel due to bad reputation for the industry.[5] Since before insurance companies have warned of the high and potentially unmanageable risks – operational and environmental – of Arctic drilling (Emmerson and Lahn, 2012).

The Arctic environment is not altogether subordinated to energy interests. But its protection is constrained by an industry whose activities linger on, to various extents, its capability to resist environmental norms. Change will not happen suddenly. Rather, as the O&G industry itself explains, this would be a very conservative industry. Just like ego and alter learning is acting through roles, the energy-environment intertwinement acts like a flexing movement between change and continuity. Thus, for as long as the global energy transition is ongoing, the Arctic region is likely to remain commercially interesting. What remains to be seen though is how much more time such energy transition would be allowed for? Should the Paris agreement be met, and global temperatures kept below 1.5 degrees' rise compared to pre-industrial levels, then emissions from carbon dioxide need to fall by 45 percent from 2010 levels by 2030, reaching 'net-zero' by 2050 (IPCC, 2018). If or when international learning encompasses a truly internalization of climate change norms, and Arctic warming specifically, things could be moving fast.

# Notes

1 See, for instance, the AC meeting Maine, 2016; Agreements on enhancing international Arctic scientific cooperation, 2017.

2 Russia has announced that it intends to increase participation in EGBCM (SAO meeting Oulu, 2017), but the expert group is yet awaiting information regarding emission projections. The state accounts for 49 percent of all black carbon emissions in the Arctic, as reported in the inventory of 2013 levels of emissions. One important reduction to make, of many, would be to reduce routine flaring, which Russia already has agreed to. Currently, Russia remains the state – in the world – with highest volumes of gas being flared annually and is responsible for almost two-thirds of Arctic flared gas (EGBCM, 2019). The U.S. is also responsible for big volumes, although the state has significantly decreased its quantities in the last few years. Overall, the U.S. projects to have its emissions of black carbon reduced by 48 percent in 2025, which would equalize a 22 percent reduction of Arctic black carbon emissions in total, compared to 2013 levels (ibid, 2019).

3 However, opposition from the American people also should have an impact, where the 'kayaktivists' mentioned elsewhere in this thesis is a symbolic illustration of how an icebreaker under the Shell logo was hindered to steer north to extraction sites. Thousands of walruses on pack ice as well as broken drilling equipment caused further delays. Most important of all is possibly the difficulties with equipment. "[I]t crushed like a beer can," U.S. Director for the Bureau of Safety and Environmental Enforcement (2012) reported, on board Shell Oil's barge the Arctic Challenger, during its sea trials before heading to the Arctic in search for oil, in 2012. What the director witnessed was how the underwater steel containment dome, safety equipment, supposed to cap a blown-out oil well, suddenly imploded.

4 Information accessed from oil and gas conferences; Arctic Oil and Gas Conference, 2014, Arctic Technology Conference, 2015, and IEA Gas & Oil Technology, at Arctic Frontiers, 2017.

5 For information accessed at oil and gas conferences, see further paragraphs.

# References

AC Ministerial Statement by AAC. (2019). *Chief Bill Erasmus' Ministerial Statement.* 11th Ministerial Meeting. Rovaniemi, Finland, May 7, 2019. (All Ministerial Statements c)

AC Ministerial Statement by Finland. (2019). *Speech by Minister Timo Soini.* 11th Ministerial Meeting. Rovaniemi, Finland, May 7, 2019. (All statement accessed via oaarchive.arctic-council.org)

AC Ministerial Statement by ICC. (2019). *Ministerial Statement.* 11th Ministerial Meeting. Rovaniemi, Finland, May 7, 2019.

AC Ministerial Statement by Norway. (2019). *Minister of Foreign Affairs of the Kingdom of Norway. Ms. Ine Eriksen Søreide.* 11th Ministerial Meeting. Rovaniemi, Finland, May 7, 2019.

AC Ministerial Statement by Saami Council. (2019). *Statement by Ms. Åsa Larsson Blind, President of the Saami Council.* 11th Ministerial Meeting. Rovaniemi, Finland, May 7, 2019.

——— (2017). *Statement by Ms. Åsa Larsson Blind, President of the Saami Council.* 10th Ministerial Meeting. Fairbanks, May 11, 2017.

Albert, M. and Vasilache, A. (2017). "Governmentality of the Arctic as an International Region," *Cooperation and Conflict,* Vol. 53:1, pp. 1–20.

Bureau of Safety and Environmental Enforcement (BSSE). (2012). *BSSE Response to KUOW FOIA.* Contributed by Ryan, John, KUOW Public Radio, November 29, 2012. Accessed on May 22, 2015, from www.documentcloud.org/documents/526393-bsee-foia-kuow-initial-response.html#document/p1.

Campbell, L. M., Corson, C., Gray, N. J., MacDonald K. I. and Brosius, P. J. (2014). "Studying Global Environmental Meetings to Understand Global Environmental Governance: Collaborative Event Ethnography at the Tenth Conference of the Parties to the Convention on Biological Diversity," *Global Environmental Politics,* Vol. 14:3, pp. 1–20.

Canada's Oil Sand Innovation Alliance. (2020). *Human Ingenuity: Reimagining Energy for Our Planet.* Accessed on February 27, 2020, from www.cosia.ca/.

Cantir, C. and Kaarbo, J. (2016). "Unpacking Ego in Role Theory: Vertical and Horizontal Role Contestation and Foreign Policy," in Cantir, C. and Kaarbo, J. (eds.). *Domestic Role Contestation, Foreign Policy, and International Relations.* London and New York: Routledge.

Carlsson, M. and Winnerstig, M. (2016). *Irreconcilable Differences. Analyzing the Deteriorating Russian-US Relations.* May 2016, FOI. Report no FOI-R – 276 –SE.

Clark, P. (2017). "Denmark's Dong Energy Shifts from Fossil Fuels to Renewables." *Financial Times,* July 17, 2017.

Emmerson, C. and Lahn, G. (2012). *Arctic Opening: Opportunity and Risk in the High North.* Produced by Lloyd's and Chatham House. London: Lloyd's.

Equinor. (2020). *Sustainability & Climate.* Accessed on February 4, 2020, from www.statoil.com/en/how-and-why/climate-change.html.

——— (2019). *Why It's Responsible to Explore the Barents Sea.* Accessed on December 12, 2019, from www.statoil.com/en/what-we-do/responsible-drilling-in-the-barents-sea.html.

European Council. (2018). *Timeline – EU Restrictive Measures in Response to the Crisis in Ukraine.* Last reviewed January 8, 2018, accessed on January 13, 2018, from www.consilium.europa.eu/en/policies/sanctions/ukraine-crisis/history-ukraine-crisis/.

Expert Group on Black Carbon and Methane (EGBCM). (2019). *Summary of Progress and Recommendations 2019.* Summary Report. The Arctic Council.

Falkner, R. (2012). "Global Environmentalism and the Greening of International Society," *International Affairs,* Vol. 88:3, pp. 503–5022.

Finnemore, M. (1996). *National Interests in International Society.* Ithaca: Cornell University Press.

Grantham Institute. (2017). *Expect the Unexpected. The Disruptive Power of Low-Carbon Technology.* Carbon Tracker Initiative, and Grantham Institute. February, 2017.

Habermas, J. (1984). *The Theory of Communicative Action. Reason and the Rationalizaton of Society,* Vol. 1. Cambridge: Polity Press.

Halinen. H. (2016). "The Arctic Council in Perspective: Moving Forward," in Heininen, L., Exner-Pirot, H. and Plouffe, J. (eds.). *Arctic Yearbook 2016. The Arctic Council: Twenty Years of Regional Cooperation and Policy-Shaping.* Iceland: Northern Research Forum.

Harnisch, S. (2015). "Role Theory and the Study of Chinese Foreign Policy," in Harnisch, S., Bersick, S. and Gottwald, J-C. (eds.). *Chinas International Roles. Challenging or Supporting International Order?* Florence: Taylor and Francis.

Heininen, L. (2015). "The Arctic Region as a Space for Trans-Disciplinary, Resilience and Peace," *Arctic and North,* Issue 21:4, pp. 69–73.

Ingebritsen, C. (2002). "Norm Entrepreneurs: Scandinavia's Role in World Politics," *Cooperation and Conflict: Journal of the Nordic International Studies Association,* Vol. 37:1, pp. 11–23.

International Energy Agency (IEA). (2017). *World Energy Outlook 2017.* Executive Summary. OECD/IEA.

Interviews. (2014–2015). Interviews conducted between December 2014 and December 2015 with state representatives and other representatives active in Arctic Council cooperation and negotiation.

IPCC. (2018). *Summary for Policymakers of IPCC Special Report on Global Warming of 1.5°C Approved by Governments.* IPCC Press release, 2018/24/PR, October 8, 2018, www.ipcc.ch/site/assets/uploads/2018/11/pr_181008_P48_spm_en.pdfKnopf, J. W. (2003). "The Importance of International Learning," *Review of International Studies,* Vol. 29:2, pp. 185–207.

Lavrov, S. (2017). Statement by Foreign Minister Sergey Lavrov at the Ministerial Meeting of the Arctic Council, Fairbanks, USA, May 11, 2017.

Letter from PPs. (2014). *Letter from PP's Regarding CO2 Emissions.* Doc 8.1. October 2014. Arctic Council Permanent Participants. Accessed from Arctic Council Open Access Repository, https://oaarchive.arctic-council.org/handle/11374/1390.

Macalister, T. (2015). "Shell Abandons Alaska Arctic Drilling." *The Guardian.* September 28, 2015.

Magnúsdóttir, G. L. and Thorhallsson, B. (2011). "The Nordic States and Agenda-Setting in the European union: How Do Small States Score?" *Icelandic Review of Politics and Administration/Stjórnmál & Stjórnsýsla,* Vol. 7:1, pp. 205–225.

Mead, G. H. (1934/1972). *Mind, Self and Society. From the Standpoint of a Social Behaviourist.* Edited and with an introduction by C. W. Morris. Chicago: University of Chicago Press.

—— (1925). "The Genesis of Self and Social Control," *International Journal of Ethics,* Vol. 35:3, pp. 251–277.

Müller, H. (2011). "Habermas Meets Role Theory: Communicative Action as Role Playing?" in Harnisch, S., Frank, C. and Maull, H. W. (eds.). *Role Theory in International Relations. Approaches and Analyses.* London and New York: Routledge.

—— (2004). "Arguing, Bargaining and All That: Communicative Action, Rationalist Theory and the Logic of Appropriateness in International Relations," *European Journal of International Relations,* Vol. 10:3, pp. 395–435.

Nicol, H. N and Heininen, L. (2014). "Human Security, the Arctic Council and Climate Change: Competition or Co-existence?" *Polar Record,* Vol 50:1, pp. 80–85.

Norwegian Oil and Gas Association (Norsk Olje og Gass). (2017). *Climate.* Published December 15, 2010, accessed November 26, 2017, from www.norskoljeoggass.no/en/Facts/The-environment/Climate/.

Pedersen, T. (2012). "Debates over the Role of the Arctic Council," *Ocean Development & International Law,* Vol. 43:2, pp. 146–156.

SAO Meeting Minutes. (2017). *Report: SAO Plenary Meeting, Oulu, Finland.* October 25–26, 2017, pp. 1–13 Accessed via oaarchive.arctic-council.org.

—— (2016). *Arctic Council Senior Arctic Officials Meeting Portland, Maine, U.S.A.* October 5–6, 2016, plenary meeting, pp. 1–25. Accessed via oaarchive. arctic-council.org.

Schober, K. (2019). "Oil Drilling in Part of Norway's Arctic Just Became a Lot Less Likely," *Arctic Today,* April 10, 2019.

Spence, J. (2017). "Is a Melting Arctic Making the Arctic Council Too Cool? Exploring the Limits to the Effectiveness of a Boundary Organization," *Review of Policy Research,* Vol. 34:6, pp. 790–811.

Staalesen, S. (2016). "Arctic Drilling Not Compatible with Arctic Drilling – Total." *The Independent Barents Observer,* May 30, 2015.

The World Bank. (2017). *World Bank Group Announcement at One Planet Summit.* Press release no. 2018/087/CGC, December 12, 2017.

Thies, C. G. (2010). "Role Theory and Foreign Policy," in Denemark, R. A. (ed.). *The International Studies Encyclopedia.* Blackwell, Blackwell Reference Online.

U.S. Office of Director of National Intelligence. (2017). *Assessing Russian Activities and Intentions in Recent US Elections.* Intelligence Community Agency, ICA 2017-01D, January 6, 2017.

Wallström, M. (2017). *Speech at the Arctic Council Ministerial Meeting in Fairbanks, Alaska, USA.* May 11, 2017.

Wilson, P. (2016). "Society, Steward or Security Actor? Three Visions of the Arctic Council," *Cooperation and Conflict,* Vol. 51:1, pp. 55–74.

World Economic Forum. (2016). *Renewable Infrastructure Investment Handbook: A Guide for Institutional Investors.* Industry Agenda. Ref 040117, December 2016. Geneva: World Economic Forum.

Wormbs, N. and Sörlin, S. (2013). "Assessing Arctic Futures. Voices, Resources, Governance," in *Mistra Arctic Futures in a Global Context.* Final Report 2011–2013.

# Epilogue
## Looking Ahead of Arctic Cooperation

### Introduction

The Arctic has changed. In the coming decades the Arctic will continue to change; some of these changes in the climate system are more difficult to assess and predict in terms of its ecological effects, whereas others are easier; an Arctic Ocean that is ice-free enough to sail during summertime will attract more economic interests – be it transportation, tourism, or oil and gas extraction. An increase in human activities also brings an increase in exposure to risks, such as accidents and pollution. Drawing on the very fact that environmental cooperation, like all cooperations, is a social act informed by social rules and expectations, this very final part of the book provides a few suggestions directed to practitioners involved in Arctic environmental cooperation. It also provides for some general notes on the Arctic Council's environmental prospectiveness into the future.

### What Is There to Learn Then?

More than twenty years have passed since the Arctic Council first gathered the Arctic states around issues on environmental protection and sustainable development. Years have passed, but no conclusion has yet been reached on how to balance the values of environmental, social, and economic character. "In general, delegates appear to want a balance between environmental protection and sustainable development, but there is not yet broad agreement about what that means in this area," a previous SAO chair summarized the AC discussion on oil and gas work (SAO meeting Fairbanks, 2016). As such, sustainable development is an illustrative example of a social world that is lacking a consensus on values. Naomi Klein has before described this lack of consensus as a battle between capitalism and the climate, arguing "What the climate needs to avoid collapse is a contradiction in humanity's use of resources; what our economic model demands to avoid collapse is unfettered expansion" (Klein, 2014:37). She concluded that where the laws of nature cannot change, it must be the rules connected to humanity's use of resources that will have to undergo change. In practice, this is an attainment that does not come around easily. Indeed, the AC case is an illustrative example of the reluctance to incorporate change at the expense of profit and economic development, and that a reversed trend concerning climate emitters takes time to establish.

Still, according to climate predictions presented, for instance, through the IPCC, there is only so much time left until the environmental costs will be too significant to bear. For the policy-maker or practitioner in the field, the question then arises of what could be done to speed up the process on environmental protection generally and climate change specifically? From the Arctic micro-perspective investigated in this thesis, a few suggestions can be provided.

*First*, any change wished for should consider actors' parallel wish for continuity. These two wishes are foundational components in any environmental progression. To have *continuity in relations* therefore is a way to ensure some stability in an unsettled future. "I guess, it is not a too shocking conclusion that personal relationships do matter," one state representative interviewed for this study laughingly said. Interaction, where delegates know each other from before or have previous experiences from the AC cooperation, seems generally to generate a more positive attitude toward negotiations and possible outcomes, than if any such experience is lacking. Although state representatives are professionals, and they are there to represent the state and not itself, they search for behavioral meaning amongst their peer negotiators: why are these other state representatives behaving the way they do? Past relational experiences are making such search easier. Another watchword for successful cooperation, connected continuity, would be *trust* – which equally increases the closer the bonds. Clearly, such measure alone cannot change a state's willingness, for instance, to regulate its energy sector, but it can decrease the likelihood of misunderstandings and falsely held expectations.

*Secondly*, the AC states ought to *expect* things from each other, things that benefit environmental protection. The flexibility in roles gives room for somewhat changed approaches, without having to abandon its own ego conception. On the contrary, alter expectations reveal new interactive rules, they reveal what kind of behavior would be considered appropriate. To do as several of the Nordic states did in the TFBCM, or as was done toward Norway in the TFOPP, and express expectations aiming high for actions on environmental protection, is a norm entrepreneurial strategy that forces others to actively reflect upon how important it is to keep a certain social position. To think about how a certain approach is affecting the current social position, how the state (i.e., *Me*) will be recognized if doing A or B, is a reflection that has to be more active the more change, i.e., new rules, states become introduced to.

*Thirdly*, to focus on *goals* of cooperation, rather than reasons for cooperation, could allow for negotiations and cooperation to put emphasis on where actors share the most commonness, and bypass some of the difficulties which follow on differentiated social realities. For instance, to establish and agree to environmental cooperation requires actors to allude to specific national justifications. If these justifications are made in relation to a specific reason of cooperation, such as health *or* climate change, actors' social life worlds are likely to cause higher amount of arguing, possibly too high amounts. Rather, to do as the TFBCM did and focus arguing and cooperation on the overarching *goal* of less pollutants, which can be related to in various ways given it is beneficial for people, animals, and nature alike, the arguing process skips some of the discourses that risk to lead cooperatively astray. Therefore, it is a strategy that appeals to change through commonness rather than difference.

*Fourthly*, in a truly theatrical vein, the AC should 'bring out the plumes' and *be bold* on Arctic environmental protection. The two negotiations studied in this book have both originated from a row of previous work conducted by the AC WGs and task forces. Since normative learning takes time as well as iterated interaction, it has been argued, environmental goals and objectives that might in the onset be experienced as too high or unrealistic will yet be perceived as more achievable and realistic as time and socialization has done its part.

To have high expectations and be bold on Arctic environmental protection would also be relevant in relation to the AC's outreach ambitions. Indeed, with the AC being a policy-shaping organization, rather than a policy-making one where sovereign states should bend for its decisions, it operates through normative influence regionally as well as globally. Environmental policy has a dramaturgical element of speech-making, and this would be symptomatic of the AC cooperation through its dedicated efforts to present the Arctic to the rest of the world. It is a staged presentation, where the AC members, through foremost speech-making on scientific findings, try to reach global adherence for the challenges facing the Arctic region. A policy-shaping organization contains not less important cooperation compared to a policy-making one, but makes a good complement. As the international understanding on the urgency of climate change mitigation currently seems to increase, the AC would have much to win to take advantage of this, to be bold on Arctic environmental protection, and thereby give climate socialization a further push forward. Given the AC's past behavior, it could be written into the AC script without causing any significant disturbances thereto, while signaling the Arctic states as committedly rising to the challenge of avoiding the worst climate effects.

### Looking Ahead

For many years, the Arctic states have stood unanimous when fronting the world outside, tied together in what could be described as 'clever' cooperation: what has made the AC cooperation impervious to state discord and conflicts boils down to a recognition of Arctic states' interdependence – in relation to environment, in relation to interests, and in relation to securing Arctic governance. The last few years has sowed dissension and doubts between the member states, possibly threatening to change the conditions for Arctic cooperation, in the direction of lowered feelings of mutuality. Political tensions have increased, and Russia's increased militarization in and out of the Arctic has here played a significant role. But there is more to it where Russia and the U.S., and China as well, are three giants engaged in what more and more has been described as an Arctic power competition.

From an environmental perspective, the U.S. has caused doubts on international climate change cooperation by withdrawing from the Paris Agreement. In the Arctic Council, the Ministerial meeting in 2019 did – for the first time in AC history – not result in a common declaration; reportedly, the U.S. would no longer support formulations of climate change. On the same occasion, Secretary of State M. Pompeo announced that the U.S. would not sign on the collective goal for reductions of black carbon, explaining how collective goals could rapidly become meaningless, even counterproductive, if even one nation fails to comply (AC Ministerial statement by the U.S., 2019). Obviously, such development in the Arctic Council is a set-back for climate cooperation. If Arctic cooperation should continue along the same lines of difference, discord would be a plausible outcome given lowered levels of trust amongst members. In relation to black carbon, the U.S. seems currently to be 'the significant other,' a stumbling block to formalized collective action on emission reductions where everybody else agrees. However, the decision not to sign on to the collective goal, should perhaps be understood less in relation to national commitments and actions and more in relation to a perceived lack of commitments by others; Russia declares itself fully supportive of collective goals but has not still – being the by far the largest emitter of black carbon – submitted national calculations on emission reductions.

Whereas the rhetoric of President Trump often has been tough on climate change, the state yet explains to 'do its part' and continues to decrease black carbon emissions in large quantities (AC Ministerial statement by the U.S., 2019; SAO meeting Rovaniemi, 2018). In a presidential meeting between the U.S. and Finland, black carbon was among the topics discussed; "we had a very good discussion," President Trump later commented, adding the states to have much in agreement (Joint Press conference, 2017). In practice, the AC cooperation on black carbon might seem more supported than its rhetoric would suggest, and therefore, the performance of the U.S. in the Arctic Council context bears resemblance with an orchestrated spectacle of strength and sovereignty. Its role as 'innate leader' has grown in proportions, where it presents itself as a leading power. Such an actor enters the international arena caring for voluntary cooperation and shared responsibilities, devolving the risk of free-riding. Such an actor, furthermore, gets not told what to do but adheres to the principle of deciding for itself. Even if those decisions in practice, after all, may end up as being rather close to what being cooperative is prescribed.

Cooperation in the Arctic Council is not likely to fall apart any time soon: there are reasons to believe that its institutions, interactive structure, and social expectations will work to lock members on to their respective cooperative role known from before, and keep relations on track. If anything, recent turbulence could spur even more committed convictions amongst the Arctic states. In a formalized statement by Finland's Minister for Foreign Affairs acting as the AC chair, where the statement would be similar in structure to the declaration going missing, the phrase "a majority of us" was used to repeatedly emphasize climate change as a fundamental challenge in need of urgent emission reductions (AC Ministerial statement by the chair, 2019). Amongst the Arctic state, a common understanding on black carbon reduction is rapidly on the rise, but it is an understanding that also presupposes action. This is something all states, the U.S. and Russia alike, have to consider and process in order to preserve their positions as cooperators in Arctic relations.

*So, will there be a future story told on Arctic preservation?* The Arctic drama is still a vivid one, with no evident direction. What Klein (2014) in the beginning of this section viewed as an impossible juxtaposition on expanded usage of natural resources, alongside climate change improvements, is a description that most Arctic states do not agree to. There is no such thing as *one paradox*, but rather *two stories* to be told, both equally relevant to the Arctic. These stories, so this book would suggest, however, seem less likely to be performed on an Arctic arena filled with major conflicts, than a cooperative one. There is much at stake in the Arctic – of economic values, but also of social and cultural value. Through the people inhabiting the region, its history, culture, and its wildlife, the Arctic states have added a dimension of uniqueness. They are Arctic – 'Northerners,' 'leaders and protectors of the top of the world.' It is a region that in a way is differentiated from the international, where Arctic dramaturgy includes story-telling on an extraordinary part of the world and of – somewhat – unique Arctic relations illustrated through the Arctic Council cooperation. Albeit speculative on the Arctic future, the thing speaking mostly in favor of enhanced environmental protection would possibly be precisely this: in their wish to stay Arctic, the Arctic states also need the support of a region being sufficiently preserved.

## References

AC Ministerial Statement by the Chair. (2019). *Statement by the Chair.* Minister for Foreign Affairs of Finland, Timo Soini, on the occasion of the eleventh Ministerial meeting of the Arctic Council. Rovaniemi, Finland, May 7, 2019. Accessed via oaarchive.arctic-council.org.

AC Ministerial Statement by the U.S. (2019) *Statement by Secretary of State Michael R. Pompeo at the Arctic Council Ministerial Meeting.* Rovaniemi, Finland, May 7, 2019. Accessed via oaarchive.arctic-council.org.

Joint Press Conference. (2017). *Remarks by President Trump and President Niinistö of Finland in Joint Press Conference.* August 28, 2017. The White House, Office of the Press Secretary.

Klein, N. (2014). *This Changes Everything. Capitalism vs. the Climate.* New York: Simon and Schuster.

SAO Meeting Minutes. (2018). *Report: SAO Plenary Meeting Rovaniemi, Finland.* November 1–2, 2018, pp. 1–25. Accessed via oaarchive.arctic-council.org.

——— (2016). *Report: SAO Plenary Meeting Fairbanks, Alaska.* March 2016, pp. 1–27. Accessed via oaarchive.arctic-council.org.

# Index